Advances in Horizon Mapping and Fault Interpretation

Kingsley Chiadikobi

Advances in Horizon Mapping and Fault Interpretation

Methodologies and Challenges in 3-D Seismic Data

Springer

Kingsley Chiadikobi
Department of Geology and Geophysics
University of Aberdeen
Aberdeen, UK

ISBN 978-3-032-23916-7 ISBN 978-3-032-23917-4 (eBook)
https://doi.org/10.1007/978-3-032-23917-4

This Springer imprint is published by the registered company Springer Nature Switzerland AG
The registered company address is: Gewerbestrasse 11, 6330 Cham, Switzerland

I dedicate this textbook to the current and next generation of geoscientists, engineers, and researchers who will interpret the subsurface with integrity and innovation.

I also dedicate this volume to mentors, supervisors, and colleagues who continue to challenge conventional thinking in structural geology and seismic interpretation.

Finally, I dedicate this textbook to my family, whose patience, encouragement, and support made the completion of this textbook possible.

May this work contribute, in its own way, to advancing Earth science for sustainable energy and the responsible development of subsurface resources.

Shalom!!!

Foreword

The global shift towards low-carbon energy increases the need to understand, characterise, and manage the subsurface. As societies expand their use of geothermal, offshore wind, subsurface storage, and carbon capture, utilisation and storage (CCUS); the demand for transparent and reproducible interpretation grows. Advanced seismic interpretation remains central to addressing this challenge.

Three-dimensional (3D) seismic reflection data provide the most detailed imaging of the Earth's shallow-to-deep subsurface at regional scales. When interpreted, seismic data reveal the structure of fault systems, stratigraphic frameworks, and fluid pathways that influence reservoir connectivity, seal integrity, and geomechanical behaviour. These elements also impact the viability and long-term safety of geothermal systems, underground hydrogen and gas storage, CO_2 sequestration, and offshore infrastructure foundations. As a result, seismic interpretation is no longer limited to hydrocarbon exploration; it has become a fundamental discipline for energy sustainability and environmental stewardship.

This textbook addresses the expanded role by providing a detailed workflow for interpreting high-resolution 3D seismic data, with a focus on structural features, uncertainty, and fault analysis. It turns interpretation from subjective guesswork into a formal process using both qualitative and quantitative methods. Documenting how interpretations are built, tested, and refined encourages reproducibility and transparency, which are essential for regulatory, energy, and environmental decisions.

The case studies and workflows show how seismic interpretation supports sustainable energy by reducing leakage risk in CCUS, guiding geothermal exploration, informing well placement, and assisting monitoring. It also helps evaluate earthquake hazards, infrastructure resilience, and environmental impacts, serving as a flexible tool for responsible subsurface management across various sectors.

This book recognises that seismic interpretation is evolving with advances in data, computation, and analytics, transforming geoscientists' work with 3D datasets. While focusing on disciplined workflows, it also lays the groundwork for integrating machine learning capabilities, such as fault detection and pattern recognition, to improve efficiency and consistency. However, automation must be grounded in a geological framework to avoid amplifying noise, bias, or artefacts.

The author have hinted that the next volume will focus on deep machine learning in 3D seismic, building on current principles and extending from human-guided interpretation to hybrid systems in which machine intelligence augments geological reasoning. These volumes combined would form a pathway from seismic interpretation through uncertainty-aware structural analysis to data-driven subsurface modelling, meeting the needs of a sustainable energy future.

This textbook aims to serve as a guide to seismic interpretation and as a reference framework for future geoscientists and engineers. By linking seismic methodology with renewable energy, environmental responsibility, and emerging machine learning technologies, it clearly positions seismic interpretation within the interdisciplinary fields of modern Earth science and sustainable development.

I admire the motivation and expertise of the author, and I highly recommend this textbook.

Aberdeen, Scotland, UK Reverend Godwin Chimara

Preface

Seismic interpretation remains one of the most powerful tools for visualising and understanding the Earth's subsurface. By mapping stratal reflectors and fault systems within three-dimensional (3D) seismic datasets, geoscientists construct geological models that inform basin analysis, tectonic reconstruction, energy exploration, and subsurface storage applications. Yet, despite technological advances in acquisition and processing, interpretation continues to face significant challenges, particularly those related to uncertainty, scale, and structural complexity.

This textbook addresses those challenges directly. It provides a systematic, reproducible, and practice-oriented workflow for interpreting high-resolution 3D seismic data from the Levantine Margin in the Eastern Mediterranean. Rather than presenting interpretation results in isolation, the book documents each analytical step, from horizon mapping and fault identification to quantitative fault scaling and structural comparison, so that readers can follow how geological understanding emerges from seismic volumes.

A fundamental theme of this volume is the clear understanding of uncertainty. Seismic interpretation involves both objective uncertainties, such as resolution limits and velocity model inaccuracies, and subjective uncertainties arising from interpretive bias. By integrating qualitative, structural–stratigraphic interpretation, and quantitative fault analysis into a unified framework, the workflow presented here reduces interpretive ambiguity and enhances reproducibility.

The Levantine Margin provides an exceptional natural type locality for testing seismic interpretation concepts due to its structural complexity. Its Plio–Quaternary and Messinian successions record gravity-driven deformation, salt-related tectonics, and complex extensional fault systems. Through detailed mapping of 364 faults across four structural domains (FM-1–FM-4), the textbook demonstrates how 3D seismic analysis reveals fault continuity, segmentation, scaling relationships, and depth-dependent variability that cannot be fully resolved in 2D sections alone.

The structure of the book reflects a progression from theory to application. It begins with foundational principles of seismic imaging and uncertainty, advances through dataset description and workflow development, and finishes in quantitative analysis and comparative studies across multiple continental margins and analogue models. This framework enables readers to build conceptual understanding while simultaneously developing practical interpretation skills.

This volume is intended for advanced undergraduate students, postgraduate researchers, academic staff, and industry professionals engaged in structural geology, basin analysis, geophysics, and subsurface modelling. It also serves emerging sectors such as carbon capture, utilisation and storage (CCUS), geothermal energy, and geomechanical modelling, where reliable structural interpretation is critical.

By actively engaging with the workflows, readers will learn how to interpret seismic data and understand why certain interpretations are more robust than others.

Aberdeen, UK Dr. Kingsley Chiadikobi

Acknowledgements I gratefully acknowledge the Tertiary Education Trust Fund (TETFund) for its funding that made this textbook possible.

I sincerely recognise the foundational contributions of Keith Aki, Paul Richards, Mikel Bacon, Rob Simm, Terry Redshaw, Alistair Brown, and Enwenode Onajite, whose established principles of seismic interpretation strengthen several workflows presented in this volume.

I further acknowledge the important work of authors whose studies on uncertainty, structural interpretation, fault growth, and 3-D seismic mapping have shaped the comparative and theoretical framework of this book.

I am particularly grateful for the structural and analogue modelling research of Ken McClay, Paul Ellis, and others, whose experimental rift-basin models provided insight into fault linkage, relay ramp breaching, and oblique extension. I also acknowledge the contributions of Adam Cawood, Kelvin Chima, and Naresh Kumar, whose work on tectonostratigraphic evolution, rift architecture, and seismic-based structural analysis contributed to the comparative synthesis presented in this text.

I gratefully acknowledge SLB (Petrel™) and Petroleum Experts (MOVE) for granting academic software licences that enabled seismic modelling and structural reconstruction. I also thank PGS, the seismic data rightsholder, for providing access to the 3D dataset for academic purposes.

I extend special appreciation to Godwin Chimara, who authored the foreword section of this book, for articulating the broader relevance of seismic interpretation to sustainable energy.

Competing Interests The author has no competing interests to declare that are relevant to the content of this manuscript.

Contents

Chapter 1
Introduction

Seismic interpretation creates three-dimensional (3-D) geological models by identifying and mapping stratal reflectors and fault systems within seismic datasets (Bacon et al., 2003; Brown, 2012; Onajite, 2014). Through this process, interpreters convert seismic volumes into meaningful representations of subsurface architecture, displayed as horizontal plan views and vertical cross sections (Carvajal et al., 2023). These visualisations form the basis for understanding basin structure, stratigraphic relationships, and tectonic evolution.

At the core of seismic interpretation is horizon mapping, which traces laterally continuous seismic reflectors that correspond to chronostratigraphic surfaces. These surfaces are usually represented as 3-D grids and contoured in two-way travel time (TWT), allowing interpreters to analyse regional geometry, thickness variations, and structural deformation (Bacon et al., 2003; Brown, 2012). Fault interpretation often accompanies this process, producing fault-pattern maps that show the distribution of displacement and structural segmentation. Despite advances in seismic acquisition and processing, both 2-D seismic lines and 3-D seismic volumes remain essential tools for subsurface interpretation in academic and industry settings.

1.1 Background

1.1.1 Challenges in Mapping Stratal Reflectors

Mapping stratal reflectors accurately remains difficult, especially in areas with poor seismic imaging or complex geology. Faulting, steeply dipping strata, and abrupt lithological changes often disrupt reflector continuity, causing uncertainty in horizon identification. In such environments, errors in tying seismic lines or volumes can spread through the interpretation, ultimately undermining the dependability of the structural model.

K. Chiadikobi, *Advances in Horizon Mapping and Fault Interpretation*,
https://doi.org/10.1007/978-3-032-23917-4_1

Interpretation uncertainty occurs when reflectors are extrapolated across zones with low signal-to-noise ratios or disrupted wavelet character. These uncertainties are not limited to local features; they can also affect regional interpretations of basin architecture, fault connectivity, and stratigraphic evolution.

1.1.2 Sources of Uncertainty in Seismic Interpretation

Uncertainty in seismic interpretation can be broadly categorised into objective and subjective components (Alcalde & Bond, 2022; Andrews et al., 2024; Bond, 2015; Tannert et al., 2007).

Objective uncertainty stems from the seismic data itself. It indicates limitations caused by acquisition geometry, processing workflows, the accuracy of the velocity model, and seismic resolution. Factors such as bandwidth, illumination, and migration artefacts restrict the smallest features that can be accurately imaged and interpreted (Andrews et al., 2024; Faleide et al., 2021; Michie et al., 2021; Pérez-Díaz et al., 2020). These constraints directly influence the confidence in detecting subtle stratigraphic variations or small-offset faults.

Subjective uncertainty, by contrast, originates from the interpreter. It reflects personal experience, conceptual assumptions, and interpretive bias, especially when dealing with complex or poorly imaged seismic features (Alcalde & Bond, 2022; Tannert et al., 2007). Divergent conceptual models, inconsistent interpretation criteria, and incomplete datasets can all increase this uncertainty. Consequently, multiple internally consistent interpretations may be produced from the same seismic volume.

Why is horizon continuity harder to establish in complex areas like fault zones or monoclines? Consider geological and data quality factors. Review a seismic section interpreted by two analysts. Which features are consistent, and where do their interpretations differ? What does this indicate about subjective uncertainty?

1.1.3 Fault Identification in 3-D Seismic Data

Fault interpretation poses distinct challenges in 3-D seismic volumes. Faults seldom appear as continuous, high-amplitude reflections. Instead, interpreters deduce their presence from indirect indicators such as reflector terminations, abrupt changes in dip, offsets, or disrupted reflector patterns. Geological features, including salt bodies, unconformities, and igneous intrusions, can further complicate interpretation by generating seismic responses that imitate fault-related discontinuities (Baudon & Cartwright, 2008; Delogkos et al., 2020; Faleide et al., 2021).

Practical fault interpretation thus depends on careful analysis of reflector geometry and termination styles. Recognising whether a disruption reflects true displacement or an alternative geological process is vital for building reliable fault models.

1.1.4 Controls on Fault Detectability

The detectability of faults in seismic data relies on two main factors:

i. Seismic data quality, including frequency content, vertical and lateral resolution, and processing fidelity, determines the minimum fault offset that can be resolved. Interpreter expertise encompasses experience, pattern recognition, and criteria for identifying structural discontinuities (Onajite, 2014).
ii. Seismic resolution fundamentally limits fault detection. Small-scale faults with displacements below the tuning thickness may remain undetected, leading to incomplete or biased structural interpretations (Dimmen et al., 2023).

1.1.5 Rethinking Interpretation Workflow

In many industry workflows, interpreters prioritise early fault mapping to speed up regional analysis. However, this method often results in poorly constrained fault geometries and internally inconsistent structural models, especially when horizons are poorly defined.

This chapter employs an alternative, systematic workflow in which horizon interpretation precedes fault delineation. By first establishing a solid stratigraphic framework, faults can be identified and constrained more confidently, thereby enhancing both structural accuracy and interpretive consistency.

> *Before mapping faults, ask: Does the seismic resolution permit identification of the displacement scale I am targeting?*

1.2 My Approach

This chapter outlines a structured method to minimise uncertainty in seismic mapping and fault interpretation by developing an integrated structural–stratigraphic framework from 3-D seismic data. The approach combines interpretive discipline with both qualitative reasoning and quantitative analysis to enhance confidence, consistency, and reproducibility. The workflow consists of three complementary components, each targeting a different source of interpretive uncertainty.

1.2.1 Structural Interpretation and Data Techniques

The first component focuses on improving the accuracy and internal consistency of structural interpretation within the 3-D seismic volume. I integrated seismic datasets into subsurface modelling software to enable interactive interpretation, three-dimensional visualisation, and iterative refinement of structural geometries.

Interpretation was performed at carefully chosen spatial scales and display resolutions to ensure that both regional trends and local structural details were adequately resolved. Vertical exaggeration was selectively applied to improve the perception of subtle fault offsets, reflector curvature, and dip variations, while avoiding geometric distortion. Throughout the workflow, I strictly followed established principles of horizon and fault mapping to preserve interpretive consistency.

Faults were identified using standardised criteria applied across seismic sections, including reflector terminations, offset patterns, and abrupt changes in dip or continuity. After interpretation, time- and depth-structure maps were generated for key horizons and fault surfaces to characterise the geometry, segmentation, and spatial organisation of major fault zones.

1.2.2 Qualitative Seismic–Stratigraphic Interpretation

The second component employs a qualitative workflow to incorporate regional and local structural patterns into a unified seismic–stratigraphic framework. Six regional 2-D seismic profiles were randomly chosen to reduce selection bias and exported at high resolution for detailed manual interpretation.

Interpretations based on the 2-D profiles were then combined with detailed horizon and fault mapping within the 3-D seismic volume. This step ensured consistency between regional observations and local structural features, enabling interpretation of stratigraphic surfaces and fault networks within a geological framework.

1.2.3 Quantitative Fault Analysis

The third component applies quantitative analysis to constrain fault interpretation and evaluate structural scaling relationships. Fault maps were analysed to extract geometric and kinematic parameters, providing an objective basis for assessing fault behaviour.

The analysis included: (i) measurements of fault displacement relative to the distance along strike, (ii) determination of fault length, (iii) estimation of maximum displacement, (iv) calculation of maximum throw, and (v) assessment of displacement gradients along each fault plane. These metrics were used to evaluate fault growth patterns, segmentation, and interactions, and to identify inconsistencies that may indicate interpretive errors or resolution constraints.

Test how vertical exaggeration influences the understanding of structural features. How does viewing a horizon in 2-D versus 3-D alter interpretation? What assumptions are made? If displacement gradients appear irregular or non-physical, first review the horizon and fault traces.

1.2.4 Integrated Interpretational Framework

Together, these three components, structural interpretation/data techniques, qualitative seismic–stratigraphic interpretation, and quantitative fault analysis, form a robust interpretative framework. By explicitly addressing both subjective and objective sources of uncertainty, the workflow improves the reliability, repeatability, and spatial resolution of seismic-based structural and stratigraphic models.

This integrated approach offers a clear pathway from raw seismic data to robust geological interpretations and can be easily applied to other 3-D seismic datasets and tectonic settings.

1.3 Textbook Layout and Chapter Structure

This textbook offers a precise and systematic workflow for seismic mapping and structural interpretation. Unlike many existing studies, which simply report results without thoroughly documenting interpretative processes, this volume clearly details each analytical step. This method allows readers to follow how uncertainties develop, propagate, and are mitigated during interpretation. By formalising this documentation, the textbook fosters reproducibility and encourages more rigorous reporting of methods in future seismic interpretation research.

The textbook consists of eleven chapters, arranged to move from theoretical foundations through methodological development to application, comparison, and synthesis.

In this chapter presents the background, highlighting the role of uncertainty in seismic interpretation and outlining the textbook's objectives and structure.

Chapter 2 introduces the datasets and tools used to interpret high-resolution three-dimensional (3-D) seismic data of the Levantine Margin. It explains the data features, resolution constraints, and visualisation techniques necessary for interpretation.

Chapter 3 concentrates on seismic imaging workflows and their implications for interpretation. It explores how the reflector character is influenced by amplitude, wavelength, acoustic impedance, scale, and resolution. The chapter also details the velocity model for the Levantine Margin dataset and clarifies the process of time-to-depth conversion within the 3-D seismic volume. Additionally, it elaborates on the interpretation methods used for Plio–Quaternary and Top Messinian strata of the Levantine Margin. Criteria for fault detection and techniques for creating fault maps are clearly documented.

Chapter 4 constructs a structural–stratigraphic framework for the Plio–Quaternary and Messinian successions, utilising regional 2-D seismic profiles. The chapter quantifies structural styles and fault polarities and examines fault–fault relationships within the extensional domain of the Levantine Margin.

Chapter 5 applies the developed workflow to a high-resolution 3-D seismic dataset from the Levantine Margin. It presents, evaluates, and interprets the resulting 3-D structural framework.

Chapters 6 and 7 offer a qualitative interpretation of faults, focusing on differences in structural styles, fault polarities, and fault attribute analysis.

Chapter 8 quantifies structural features using statistical techniques for fault analysis, including Power-Law Relationships in Fault Scaling, Regression Analysis in Log–Log Space, Confidence Intervals in Scaling Analysis, Displacement Gradient, Fault Displacement, and Fault Growth.

Chapter 9 compares the results from the Levantine Margin with three additional case studies: (i) offshore Western Niger Delta Basin, (ii) the Orphan Basin–Flemish Pass on the Newfoundland Margin, eastern Canada, and (iii) the Hanna Trough and North Chukchi Basin on the U.S. Chukchi Shelf, Arctic Alaska. These comparisons are based on reproduced and re-analysed seismic sections and maps from recent peer-reviewed studies, enabling an evaluation of similarities and contrasts across different rifted-margin settings.

Chapter 10 expands the comparative analysis by examining structural styles in the Levantine Margin against analogue rift experiments using sandbox models of continental margin extension. This chapter connects natural seismic observations with experimental results, emphasising scale effects and deformation mechanisms.

Chapter 11 synthesises the main findings of the research. It summarises key insights into seismic interpretation uncertainty, fault system development, and structural–stratigraphic evolution, and outlines implications for future seismic interpretation workflows and rift basin studies.

Learning Objectives

After completing this chapter, readers will be able to:

1. Explain the role of seismic interpretation in analysing subsurface structures and applications related to energy transition.

2. Differentiate between objective and subjective uncertainty sources in seismic interpretation.
3. Evaluate how interpreter bias and data limitations influence structural models.
4. Justify the need for systematic, reproducible interpretation workflows.
5. Recognise the limitations of interpretations in complex geological settings.

References

Alcalde, J., & Bond, C. E. (2022). *Subjective uncertainty and biases: The impact on seismic data interpretation* (pp. 103–123). Elsevier.

Alcalde, J., Bond, C. E., Johnson, G., Butler, R. W. H., Cooper, M. A., & Ellis, J. F. (2017). The importance of structural model availability on seismic interpretation. *Journal of Structural Geology*, *97*, 161–171. https://doi.org/10.1016/j.jsg.2017.03.003.

Andrews, B. J., Mildon, Z. K., Jackson, C. A., & Bond, C. E. (2024). Quantifying fault interpretation uncertainties and their impact on fault seal and seismic hazard analysis. *Journal of Structural Geology, 184*, Article 105158. https://doi.org/10.1016/j.jsg.2024.105158

Bacon, M., Simm, R., & Redshaw, T. (2003). *3-D Seismic Interpretation.* Cambridge University Press.

Baudon, C., & Cartwright, J. (2008). Early-stage evolution of growth faults: 3-D seismic insights from the Levant Basin, Eastern Mediterranean. *Journal of Structural Geology, 30*(7), 888–898. https://doi.org/10.1016/j.jsg.2008.02.019

Bond, C. E. (2015). Uncertainty in structural interpretation: Lessons to be learnt. *Journal of Structural Geology, 74*, 185–200. https://doi.org/10.1016/j.jsg.2015.03.003

Bond, C. E., Gibbs, A. D., Shipton, Z. K., & Jones, S. (2007). What do you think this is? "Conceptual uncertainty" in geoscience interpretation. *GSA Today (Geol. Soc. Am.) 17*, 4. https://doi.org/10.1130/GSAT01711A.1.

Brown, A.R. (2012). *Interpretation of Three-Dimensional Seismic Data* (6th ed.). AAPG Memoir/SEG Investigations in Geophysics, No. 9.

Carvajal, F. R., Butler, R. W., & Bond, C. E. (2023). Mapping faults in 3-D seismic data–why the method matters. *Journal of Structural Geology, 177*, Article 104976. https://doi.org/10.1016/j.jsg.2023.104976

Cox, D. R., Knutz, P. C., Campbell, D. C., Hopper, J. R., Newton, A. M., Huuse, M., & Gohl, K. (2020). Geohazard detection using 3-D seismic data to enhance offshore scientific drilling site selection. *Scientific Drilling, 28*, 1–27.

Delogkos, E., Saqab, M. M., Walsh, J. J., Roche, V., & Childs, C. (2020). Throw variations and strain partitioning are associated with fault-bend folding along normal faults. *Solid Earth, 11*, 935–945. https://doi.org/10.5194/se-11-935-2020

Dimmen, V., Rotevatn, A., Lecomte, I. (2023). Imaging of small-scale faults in seismic reflection data: insights from seismic modelling of faults in outcrop. *Marine and Petroleum Geology*, *147*. https://doi.org/10.1016/j.marpetgeo.2022.105980.

Faleide, T.S., Braathen, A., Lecomte, I., Mulrooney, M.J., Midtkandal, I., Bugge, A.J., Planke, S. (2021). Impacts of seismic resolution on fault interpretation: insights from seismic modelling. *Tectonophysics,* 816. https://doi.org/10.1016/j.Tecto.2021.229008.

Iacopini, D., Butler, R. W. H., Purves, S., McArdle, N., & De Freslon, N. (2016). Exploring the seismic expression of fault zones in 3-D seismic volumes. *Journal of Structural Geology, 89*, 54–73. https://doi.org/10.1016/j.jsg.2016.05.005

Magee, C., Love, V., Fayez, K., Andrews, B., Rivas-Dorado, S., Jackson, C., Orlov, C., & Bramham, E. 2023. Quantifying dyke-induced graben and dyke structure using 3-D seismic reflection data and the role of interpretation bias. Teκ τ https://doi.org/10.55575/tektonika2023.1.2.25.

Michie, E. A. H., Mulrooney, M. J., & Braathen, A. (2021). Fault interpretation uncertainties using seismic data, and the effects on fault seal analysis: A case study from the Horda platform, with Implications for CO2 storage. Preprint. Tectonic plate interactions, magma genesis, and lithosphere deformation at all scales/Structural geology and tectonics, rock physics, experimental deformation/Structural geology. https://doi.org/10.5194/se-2021-23.

Misra, A. A., & Mukherjee, S. (2018). *Atlas of Structural Geological Interpretation from Seismic Images*. John Wiley & Sons.

Onajite, E. J. (2014). *Seismic Data Analysis Techniques in Hydrocarbon Exploration* (pp. 198–202). Elsevier.

Pérez-Díaz, L., Alcalde, J., & Bond, C. E. (2020). Introduction: Handling uncertainty in the geosciences: Identification, mitigation, and communication. *Solid Earth, 11*, 889–897. https://doi.org/10.5194/se-11-889-2020

Steventon, M. J., Jackson, C.A.-L., Hall, M., Ireland, M. T., Munafo, M., & Roberts, K. J. (2022). Reproducibility in subsurface geoscience. *Earth Science, Systems and Society, 2*, 10051. https://doi.org/10.3389/esss.2022.10051

Tannert, C., Elvers, H.-D., & Jandrig, B. (2007). The ethics of uncertainty. *EMBO Reports, 8*, 885–974. https://doi.org/10.1038/sj.embor.7401072

Chapter 2
Understanding the 3D Seismic Data Set

Abstract This chapter introduces the fundamentals of working with a large, high-resolution three-dimensional (3D) seismic reflection dataset from the Levantine Margin. It guides the reader through a complete, practice-oriented interpretation workflow. You are taken step by step from seismic data acquisition and processing to visualisation, velocity modelling, and structural analysis, showing how geological insight emerges from raw seismic volumes. The regional tectonic and stratigraphic framework of the Levant Margin is outlined to give essential background, emphasising Messinian evaporite deposition, post-Messinian thin-skinned extension, and salt-controlled deformation. Core principles of seismic imaging are explained, including reflector creation, scale and resolution limits, vertical exaggeration, and time-to-depth conversion using velocity models. Practical examples demonstrate how interpretation choices affect fault geometry, horizon mapping, and structural understanding. The chapter also highlights common limitations of 3D seismic data, especially in salt-affected settings, and promotes critical evaluation of uncertainty. By actively engaging with the dataset and workflows presented, you will develop a strong foundation for interpreting complex extensional systems and for applying these skills to other rifted and salt-influenced continental margins.

Keywords Seismic visualisation · Velocity modelling · Seismic resolution · Vertical exaggeration · Acoustic impedance · Geological uncertainty

2.1 Introduction

In this chapter, I employ a three-dimensional (3D) high-resolution seismic reflection dataset from the Levantine Margin. This dataset was acquired and processed by Petroleum Geo-Services (PGS Geophysical, 2013a, 2013b). Throughout this chapter, I utilise it to demonstrate how large-scale seismic data can be interpreted with modern geoscientific software, including Petrel, Landmark DSG, Paleoscan, OpendTect, Paradigm, and MOVE.

K. Chiadikobi, *Advances in Horizon Mapping and Fault Interpretation*,
https://doi.org/10.1007/978-3-032-23917-4_2

To follow the discussion, imagine you are working with this dataset. You begin with the raw seismic volumes and gradually use tools for interpretation, mapping, and structural analysis. This practical approach illustrates how geoscientists transition from data collection to geological understanding.

The chapter begins by describing the location, formation history, tectonic evolution, and geological setting of the Levantine Margin in the Eastern Mediterranean. These aspects have been covered in many classical and modern studies (e.g., Garfunkel, 1981; Gradmann et al., 2005; Gvirtzman et al., 2013, 2017; Hsü et al., 1973; Walley, 1998). Building on this literature, I focus on the structural features that make the region an ideal analogue for rift systems.

Earlier work showed that these structural features can be interpreted in 2D seismic cross-sections (Clark & Cartwright, 2009; Nwosu, 2013). In this chapter, I build on that understanding by exploring fault interactions in three dimensions, emphasising their role in governing deformation within the extensional domain.

Finally, I introduce the fundamental principles of seismic imaging. Reflectors originate from contrasts in acoustic impedance, amplitude, and wavelength across subsurface layers. Understanding how scale, resolution, and velocity models influence these reflectors is essential. In particular, the velocity model allows conversion from the time domain (two-way travel time, TWT) to depth (metres), which is crucial for accurate structural interpretation. I also emphasise the inherent limitations of 3D high-resolution seismic reflection data to facilitate critical evaluation of the dataset's strengths and uncertainties.

By the end of this chapter, you should be able to appreciate not only the technical workflow for handling large 3D seismic volumes but also the geological insights these datasets provide when integrated with structural and tectonic frameworks.

2.2 Location and Formation

It is located offshore within the Levant Margin of the eastern Mediterranean. It is bounded to the north by the Cyprus Arc and the Larnaca Thrust Zone, to the west by the Eratosthenes Seamount and the Herodotus Basin, to the southwest by the Nile Delta deep-sea cone, and to the east by the eastern Mediterranean coastline (Tibor & Ben-Avraham, 2005; Gardosh et al., 2006; Roberts & Peace, 2007) (see Fig. 2.1). The Levant Basin itself is Mesozoic in age (Elfassi et al., 2019).

The formation of the basin is closely associated with the Messinian Salinity Crisis. During the Late Miocene (~5.5 Ma; see Fig. 2.2), the Atlantic–Mediterranean gateways narrowed, isolating the Mediterranean from the open Atlantic. This restriction caused extreme evaporation and widespread evaporite deposition (Elfassi et al., 2019; Hsü et al., 1973; Ryan et al., 1973). In the Levant Basin, evaporite layers reach nearly 2 km in the deepest depocentres but become thinner landward, tapering to only a few tens of metres beneath the present-day continental slope (Bertoni & Cartwright, 2006; Gradmann et al., 2005; Ryan & Cita, 1978).

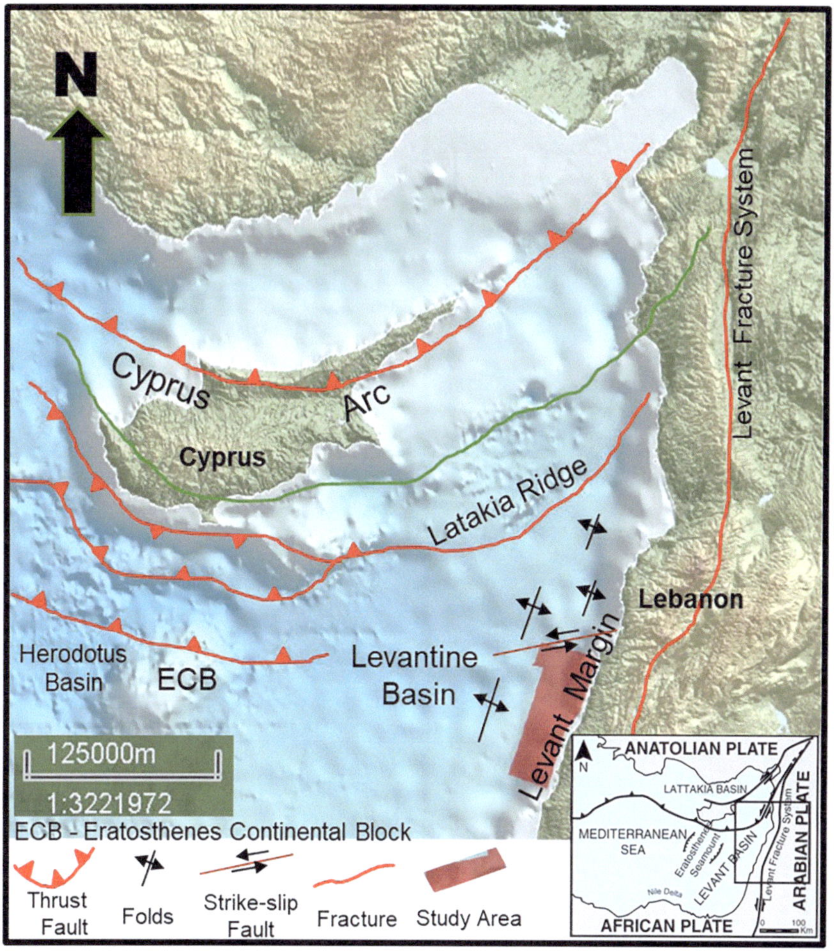

Fig. 2.1 Location map displaying topography overlaid with key geological structures of the region, Levantine Margin, and the bathymetry of the eastern Mediterranean (Modified after Ghalayini et al., 2018; Skiple et al., 2012)

The crisis abruptly ended when the Strait of Gibraltar reopened, allowing the Atlantic Ocean to refill the Mediterranean Basin in a rapid geological event. This flooding marked the start of Plio–Quaternary sedimentation, which covered the basin with siliciclastic and hemipelagic deposits (Clauzon et al., 1996; Garcia-Castellanos et al., 2020). A notable consequence of the salinity crisis was the deep erosion of submarine canyons along the continental margin. Carved during sea-level drawdown, these canyons were later filled with Messinian evaporites and subsequently buried beneath Plio–Quaternary deposits (Gvirtzman and Buchbinder, 1978; Druckman et al., 1995).

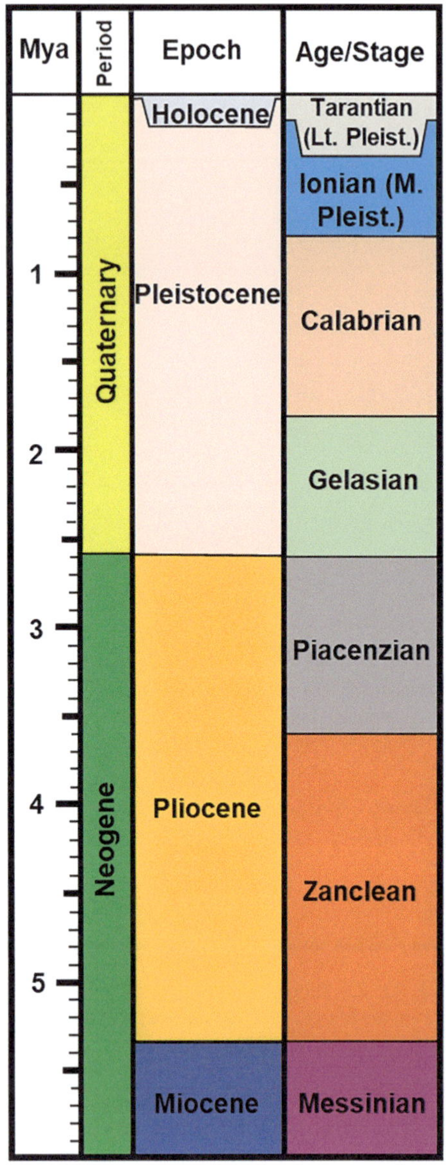

Fig. 2.2 Chronostratigraphic chart of the Plio-Quaternary and Messinian sections of the Levantine Margin (Modified after Gradstein et al., 2012)

2.3 Tectonic and Geologic Settings

The tectonic evolution of the Levant Margin began in the Palaeozoic, following the breakup of Gondwana. By the Late Cretaceous, the margin was situated between the converging Afro-Arabian and Eurasian plates (Freund, 1975; Garfunkel, 1981,

1998; Ben-Avraham et al., 1978, 2002). This plate interaction established the Levant Margin as a deep marine basin along the northern Lebanese coast.

During the Late Miocene, the Levant Fracture System (LFS) developed in conjunction with the opening of the Gulf of Aden and the Red Sea. This tectonic event caused the separation of the Arabian Plate from the African Plate (Beydoun, 1999; Elfassi et al., 2019; Gardosh et al., 2006; Hempton, 1987). The LFS remains a major structural boundary that impacts regional tectonics and sedimentation patterns.

A key focus is the up-dip extensional domain on the present-day continental slope. This zone contains syn-sedimentary extensional faults of various throw, dip, and length, spanning approximately 10–25 km wide (see Fig. 2.3). The margin's outer shelf and slope are mostly formed by thin-skinned extensional structures, including grabens, slumps, rotated blocks, and listric normal growth faults (Ghalayini et al., 2018).

The most notable features are listric and growth-faulted anticlines, which formed due to lateral displacement of the Messinian evaporites and their overburden (Elfassi et al., 2019; Ghalayini et al., 2018). These structures demonstrate how multilayer mechanical stratigraphy develops when mobile evaporite facies detach at various stratigraphic levels (Cartwright & Jackson, 2008; Garfunkel, 1984; Gradmann et al., 2005).

Previous studies indicate that regional tilting and differential sediment loading along the Levant Margin initiated a post-Messinian gravitational collapse system, which subsequently deformed the up-dip extensional domain of the Levant Basin (Cartwright & Jackson, 2008; see Fig. 2.4). The maximum principal stress direction

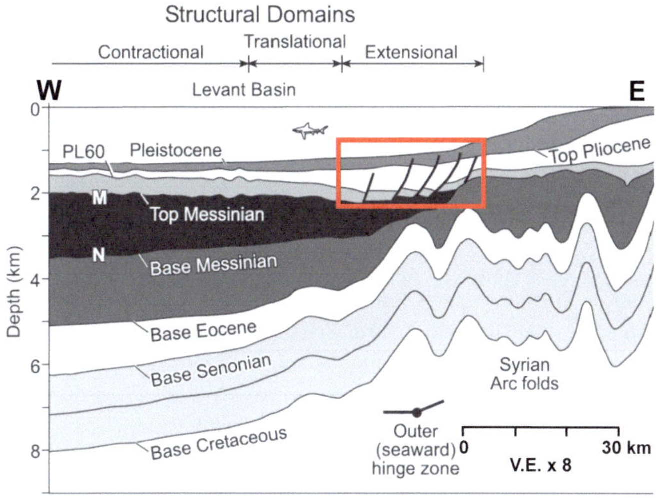

Fig. 2.3 The regional cross-section displays the main stratigraphic units, offshore thin-skinned structural zones, and inner and outer crustal hinge zones (Adapted after Cartwright & Jackson, 2008)

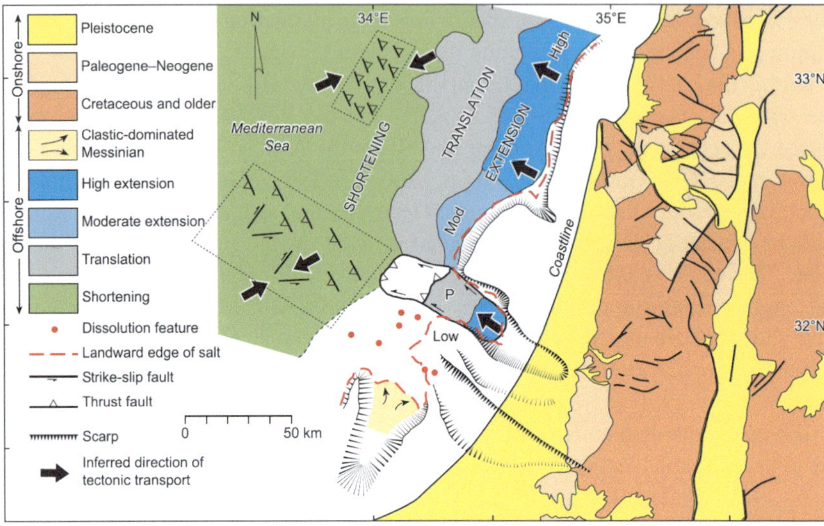

Fig. 2.4 Map of the structural characteristics and domains of the Levant margin. The Messinian salt's up-dip edge (red dashes) and the up-dip limit of extension are tightly correlated (blue areas). The frontal scarp (hachured) in the north also follows these features but diverges significantly in the south, where it enters and merges with major pre-Messinian canyons. P marks the undersea landslide at Palmahim. The three geological units, whose ages correspond to increasing elevation, demonstrate the extent of onshore exhumation (Adapted after Bartov, 1979)

related to thin-skinned deformation ranges from 100° to 120° (Cartwright & Jackson, 2008).

In the northern and central sectors, extensional domains appear relatively linear. Toward the southern margin, however, strain decreases from high to low, reflecting changes in structural style (Cartwright & Jackson, 2008). Along almost the entire continental margin, the up-dip limit of mobile Messinian evaporites coincides with the landward boundary of the extensional domain, thereby emphasising the strong structural control exerted by salt distribution (Cartwright & Jackson, 2008).

The Levantine Margin exhibits salt mobility that influences thin-skinned extension, fault interaction, and structural division. Similar processes occur along the margins of the Gulf of Mexico and the South Atlantic, where extensive evaporites shape fault structures and hydrocarbon traps. Lessons from Lebanon, such as salt thickness, differential loading, and velocity modelling, can guide exploration and tectonic studies in similar salt-affected basins.

2.4 The Levantine Margin

It involves a 3D seismic survey covering roughly 5,372 km^2. Petroleum Geo-Services (PGS) acquired these datasets, and PGS data processing centres converted them into high-resolution seismic volumes (PGS Geophysical, 2013a, 2013b). Interpretations are performed in laboratories or on workstations (see Fig. 2.5) using geological and geophysical software packages such as Petrel, Landmark DSG, Paleoscan, OpendTect, Paradigm, and MOVE.

To extract meaningful geological information, a combination of techniques was employed: integrating 3D seismic data into simulation software; visualising seismic volumes; applying vertical exaggeration; performing velocity modelling; and conducting resolution analysis. These workflows collectively ensured that the seismic dataset could be analysed with both technical precision and geological insight.

Fig. 2.5 Seismic interpretation workstations (SeisLab) within the Department of Geology and Geophysics

2.4.1 Integration of 3D Seismic Data into Simulation Software

A structured workflow was established to import and integrate the 3-D seismic volume into interpretation software, enabling improved subsurface visualisation. The dataset, supplied in SEG-Y format, was georeferenced to a global coordinate system to ensure precise spatial positioning. For confidentiality reasons, the exact coordinate reference system is not revealed.

Due to the large dataset size, a data realisation technique was used. This step decreased file volume while maintaining essential structural and stratigraphic features. Benchmark tests comparing the original and reduced datasets confirmed that the streamlined volumes kept the accuracy needed for fault and horizon interpretation. To further protect interpretive quality, an iterative refinement process was applied, ensuring that the reduced volumes remained true to the geological features of interest.

This integration strategy provided two main advantages: (1) effective data management and processing efficiency, and (2) a dependable platform for structural mapping, reflector interpretation, and geological model development across the seismic volume.

2.4.2 Visualisation of the 3D Seismic Data

The 3D seismic volume was visualised in multiple orientations to facilitate comprehensive interpretation. These orientations included the regional 2D seismic profile (see Fig. 2.6a), two vertical planes (inline and crossline sections; see Fig. 2.6b), and a horizontal plane represented by time slices (Z-line; see Fig. 2.6c).

All seismic data were presented in two-way travel time (TWT), measured in seconds. TWT indicates the time it takes for seismic energy to travel through the subsurface, reflect off geological interfaces, and return to the receivers. In this format, seismic traces record variations in acoustic impedance, producing characteristic waveforms characterised by amplitude and wavelength.

Each seismic trace consists of alternating peaks and troughs (see Fig. 2.7a), which serve as the fundamental units of seismic interpretation (Bacon et al., 2007; Brown, 2012; Onajite, 2014). Colour-coded reflectors improved visual clarity: blue was used for peaks and red for troughs (see Fig. 2.7b). This colour scheme made it easier to identify subtle reflector terminations and fault offsets within the seismic volume.

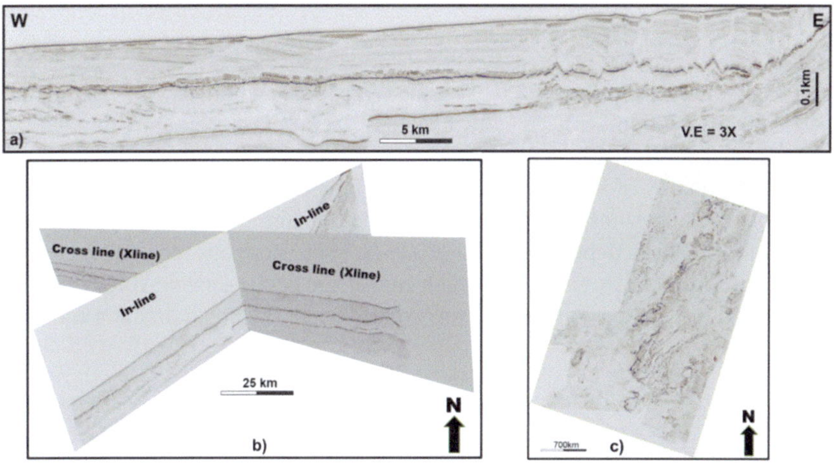

Fig. 2.6 Visualisation of seismic data. **a** 2-D regional profile from the Levantine Margin dataset. **b** 3-D seismic sections of the sliced vertical dimensions (inline and crossline) from the Levantine Margin dataset. **c** Time slice from the Levantine Margin dataset

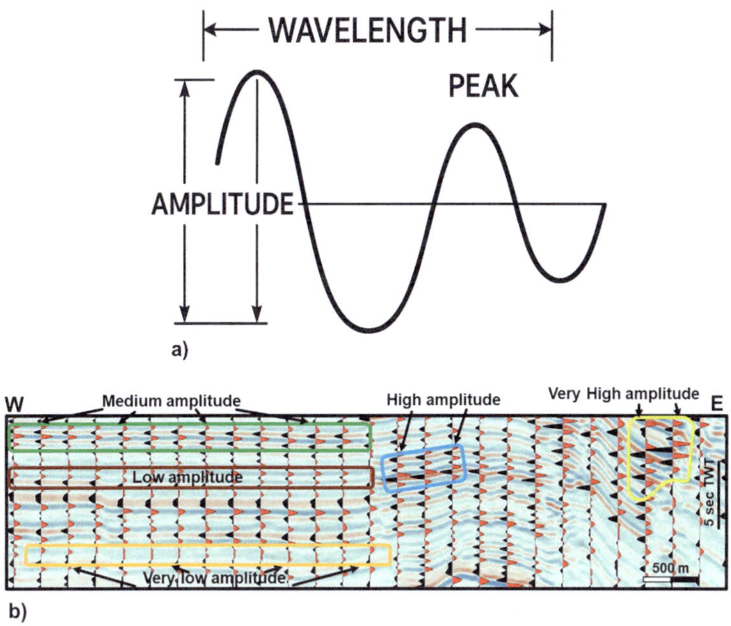

Fig. 2.7 **a** Schematic diagram showing the components of a seismic trace; **b** Seismic section with a trace showing various amplitude intensities and the correlation of the wiggles with the reflectors

2.4.3 Velocity Model

The velocity model was developed using the HyperBeam workflow, which combines initial model creation with velocity updates based on wavelet-shift tomography (HyperTomo) (PGS, 2013). This method ensured that velocity enhancements closely matched the quality of seismic data and imaging precision.

Key seismic horizons, measured in the time domain (ms), served as reference markers for pre-stack depth migration (PSDM). These horizons included regionally important reflectors that guided the depth conversion of the seismic volume.

An initial interval velocity model was created using the linear gradient equation.

$$V = V_0 + kz$$

where:

- V_0 is the initial velocity at the top of the layer,
- k is the velocity gradient (s^{-1}),
- z is depth.

Both V_0 and k were obtained from velocity scans and flatness analysis of seismic gathers. The resulting model, provided by Petroleum Geo-Services (PGS), was used to convert the processed pre-stack time-migrated (PSTM) dataset from time (TWT) to depth (m). A summary of the initial velocity parameters for the main layers is provided (see Table 2.1).

For the target interval between the seabed and the base of the Pliocene, a constant velocity of 1975 m/s was applied ($k = 0$). This streamlined the time–depth conversion and helped prevent potential overfitting linked to gradient-based models. Conversion from two-way travel time (TWT) to depth was calculated using the standard formula:

$$z = \frac{V.t}{2}$$

For example: At 1.8 s, the depth z is:

$$z = \frac{1975 \times 1.8}{2} = 1777.5\text{m}$$

This velocity model was incorporated into the interpretation software, facilitating direct time-to-depth conversion of seismic reflectors. Using predefined TWT surface

Table 2.1 Initial velocity parameters for the main stratigraphic intervals of the Levantine Margin

Layer	V_0 (m/s)	k
Sea surface—Seabed	1525	0
Seabed—Base Pliocene	1975	0
Base Pliocene—Base Salt	3500	1.2

Table 2.2 Time–depth conversion constant velocity (1975 m/s)

Time (sec)	Depth (m)
1.8	1777.5
1.9	1876.25
2	1975
2.1	2073.75
2.2	2172.5
2.3	2271.25
2.4	2370
2.5	2468.75
2.6	2567.5
2.7	2666.25
2.8	2765

maps and corresponding velocity inputs, depth-structure maps were generated for all interpreted horizons. Table 2.2 presents the constant velocity time–depth conversion for the interval between 1.8 and 2.8 s TWT, corresponding to depths of 1777.5–2765 m (see Fig. 2.8a).

A regression analysis of the TWT–depth relationship confirmed a perfectly linear correlation.

$$\text{Depth (m)} = 987.5 \times \text{Time (s)}$$

This linear function, with a slope of 987.5 m/s and a zero intercept, yielded a coefficient of determination (R^2) = 1.000 and a standard error of the estimate = 0.000. These results validate the reliability of the constant-velocity approach for this stratigraphic interval.

The model supports robust time-to-depth conversion (see Fig. 2.8b) and is particularly effective for structural interpretation, stratigraphic mapping, and fault analysis, where velocity variations are minimal.

A constant-velocity model is appropriate for this interval because the stratigraphy between the seabed and the base of the Pliocene is relatively homogeneous, with minimal lateral or vertical velocity variations. In such settings, assuming a single average velocity provides a reliable approximation for time-to-depth conversion without introducing significant error.

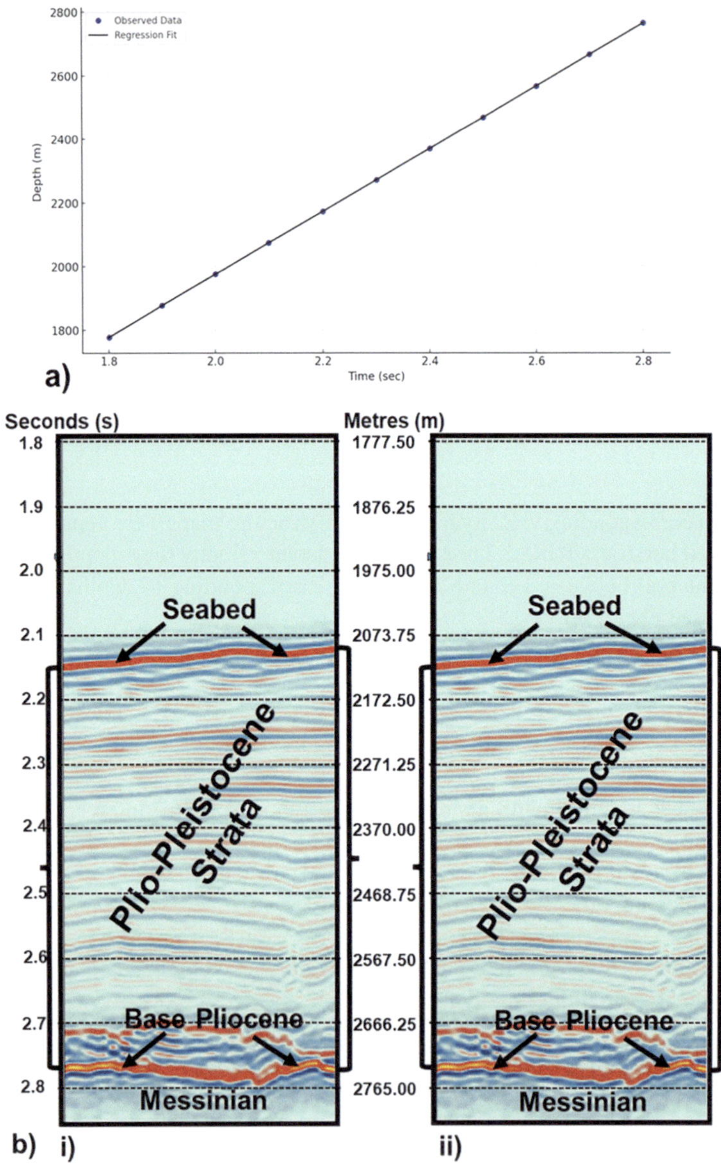

Fig. 2.8 Time-to-depth calibration and seismic section annotation using a regression-based velocity model. **a** Depth conversion plot showing a linear regression fit (black line) between observed two-way seismic travel time (s) and corresponding depth (m), indicating a consistent velocity model with high correlation. Blue dots represent observed time-depth pairs derived from key reflectors. **b** Calibrated seismic section annotated with the corresponding time (left) and depth (right) axes. Horizons and reflectors are accurately correlated using the regression-derived depth values, enabling structural and stratigraphic interpretation in both the time and depth domains

2.4.4 *Scale and Resolution*

The scale of a 3D high-resolution seismic volume is determined by its horizontal (XY) and vertical (Z) dimensions. Scale influences how geological features are visualised and interpreted, but it does not modify the inherent resolving power of the seismic data (Bacon et al., 2007; Brown, 2012; Onajite, 2014).

During interpretation, primary reflectors and faults were generally mapped at display scales of 1:100,000 and 1:10,000, respectively (see Fig. 2.9a–e). In structurally complex areas, a more detailed scale of 1:5,000 was used to improve visibility of subtle reflector geometries and minor fault traces (see Fig. 2.9f). These scales were chosen based on structural complexity, interpretive goals, and the need for visual clarity. Importantly, while finer scales magnify detail, they do not enhance the dataset's inherent resolution.

a. Seismic Resolution

Seismic resolution is the inherent ability of a dataset to distinguish two closely spaced reflection events, either vertically (in time) or laterally (in space). It is controlled by wavefield physics, acquisition geometry, and processing techniques (SEG Wiki, 2024).

Following the Rayleigh criterion, vertical resolution (Rv) is estimated using the quarter-wavelength rule.

$$\mathrm{Rv} \approx \frac{\lambda}{4}, \text{ where } \lambda = \frac{v}{f},$$

where λ is wavelength, v is seismic velocity, and f is the dominant frequency.

For this dataset, with v =1975 m/s and f =40 Hz:

$$\lambda = \frac{1975}{40} = 49.40, \mathrm{R_v} = \frac{49.40}{4} \approx 12.35 \text{ m}$$

This suggests that two reflectors need to be at least 12.35 m apart vertically in order to be distinguished as separate reflections.

b. Lateral resolution

Lateral resolution, governed by the Fresnel zone diameter, improves considerably with pre-stack time migration (PSTM). Migration collapses diffractions, enhances reflector continuity, and enhances imaging in structurally complex areas (e.g., Yilmaz, 2001). PSTM parameters were optimised to enhance lateral resolution. Additional processing techniques, including spectral broadening and deconvolution, increased the frequency bandwidth, sharpened the seismic wavelet, and improved vertical resolution, thus enabling better imaging of thin beds and minor faults.

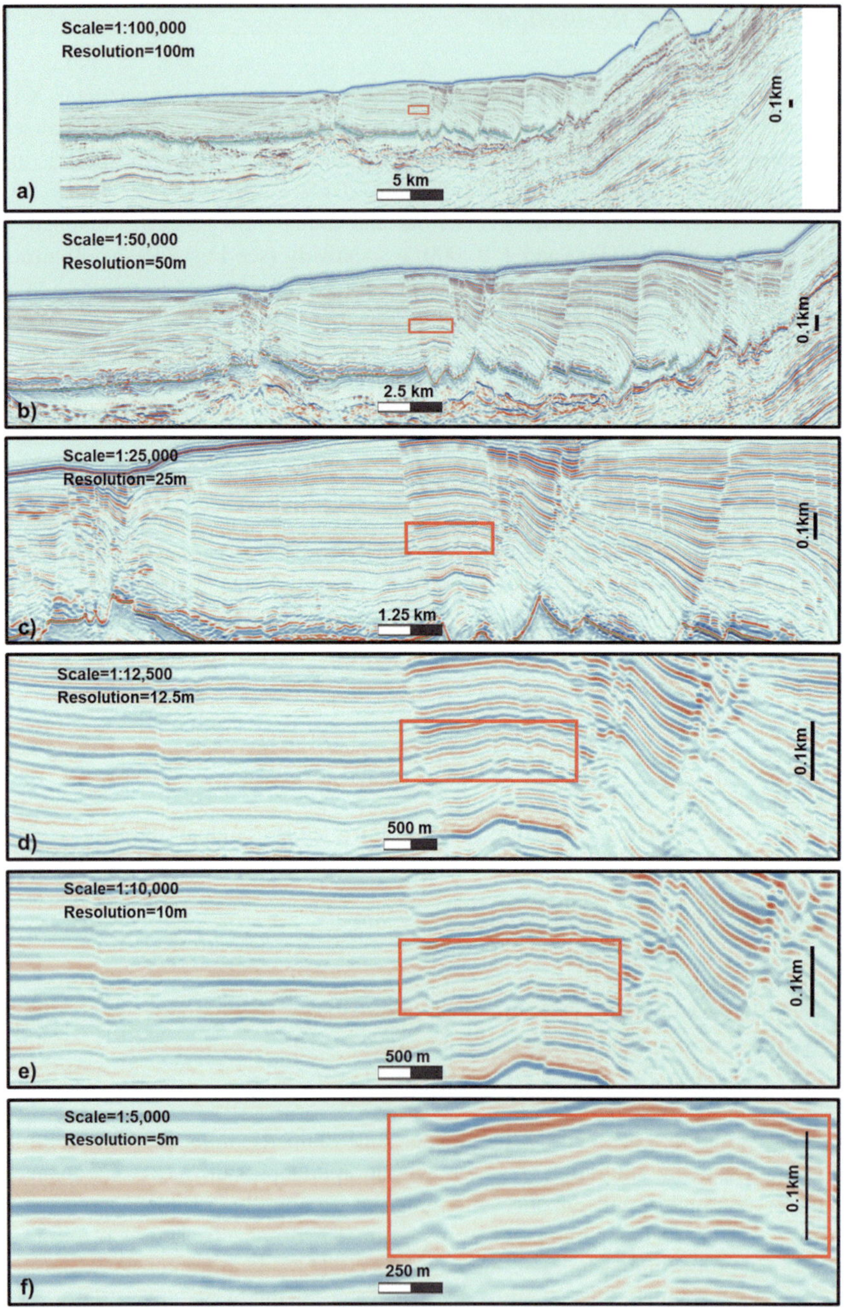

Fig. 2.9 Relationship between scale and resolution on a seismic section. **a** 1:100,000/100 m, **b** 1:50,000/50 m; **c** 1:25,000/25 m, **d** 1:12,500/12.5 m, **e** 1:10,000/10 m, and **f** 1:5,000/5 m. Increasing resolution improves the clarity of internal reflector geometries, continuity, and deformation patterns. Red boxes highlight the area of focus, displaying the same stratigraphic interval across various scales and revealing finer details of fault displacement, reflector terminations, and layer architecture that are essential for structural interpretation and reservoir characterisation

Table 2.3 Relationship between scale and resolution

s/n	Scale (RF)	Detectable size resolution (m)	Raster resolution (m)
1	1:100,000	100	50
2	1:50,000	50	25
3	1:25,000	25	12.5
4	1:12,500	12.5	6.25
5	1:10,000	10	5
6	1:5,000	5	2.5

Practical Outcomes

Structural features, such as faults, were identified using seismic discontinuities, acoustic impedance contrasts, and velocity variations, which were calibrated against the velocity models (see Sect. 4.2). Although apparent resolutions from 100 to 5 m could be achieved through scale magnification (see Fig. 2.9a–f), actual detectability remained limited by the physical resolution constraints of the data.

Selected 2D seismic sections were exported at 300 dpi using high-resolution graphics software. Although this ensured clarity in print, it did not alter seismic resolution, which is fundamentally controlled by acquisition design, frequency content, rock properties, and processing parameters. By carefully balancing display scale with data resolution, it achieved accurate delineation of fault geometries, including both low- and high-angle fault planes across multiple structural domains. The relationship between scale and resolution is summarised in Table 2.3.

2.4.5 *Vertical Exaggeration in Seismic Interpretation*

Vertical exaggeration (VE) is a widely used technique in seismic interpretation that improves the visibility of subsurface structures by stretching the vertical scale more than the horizontal scale (Bacon et al., 2003; Stewart, 2011, 2012; Donnellan et al., 2017). By increasing vertical relief, VE emphasises reflector offsets, subtle dips, and small-scale structural features that might otherwise be missed.

However, excessive VE causes distortions. Fault dips may appear steeper or shallower than their true geometries, reflector thicknesses may be misrepresented, and depth relationships between horizons may become skewed (Hatton et al., 2003; Hubral & Tygel, 1997; Stone, 1991). Furthermore, exaggerated vertical scales can overemphasise shallow structures while visually compressing deeper ones, complicating structural reconstructions and misleading depth-to-horizon estimates (Müller & Müller, 2020; Sheriff & Geldart, 1995; Wu & Wu, 2017).

In practice, VE values usually range from 2 × to 4 ×. Values exceeding 3 × are regarded as extreme and should be approached with caution, as they can distort geological gradients (Bacon et al., 2003; Stewart, 2011).

In the Levantine Margin dataset, VE values of 1 ×, 3 ×, and 5 × were utilised according to interpretation objectives:

- 1 × (true scale): Provided geometrically accurate representations (see Fig. 2.10a), suitable for fault dip and depth analyses.
- 3 × (moderate exaggeration): Improved visibility of fault throw variations and subtle reflector geometries on 2D profiles (see Fig. 2.10b).
- 5 × (strong exaggeration): Used in 3D volumes for horizon and fault mapping, highlighting reflector offsets and subtle dip variations (see Fig. 2.10c).

To ensure interpretive accuracy, distortions caused by 5 × magnification were cross-checked with displays at 1 × and 3 ×. When discrepancies occurred, geometric

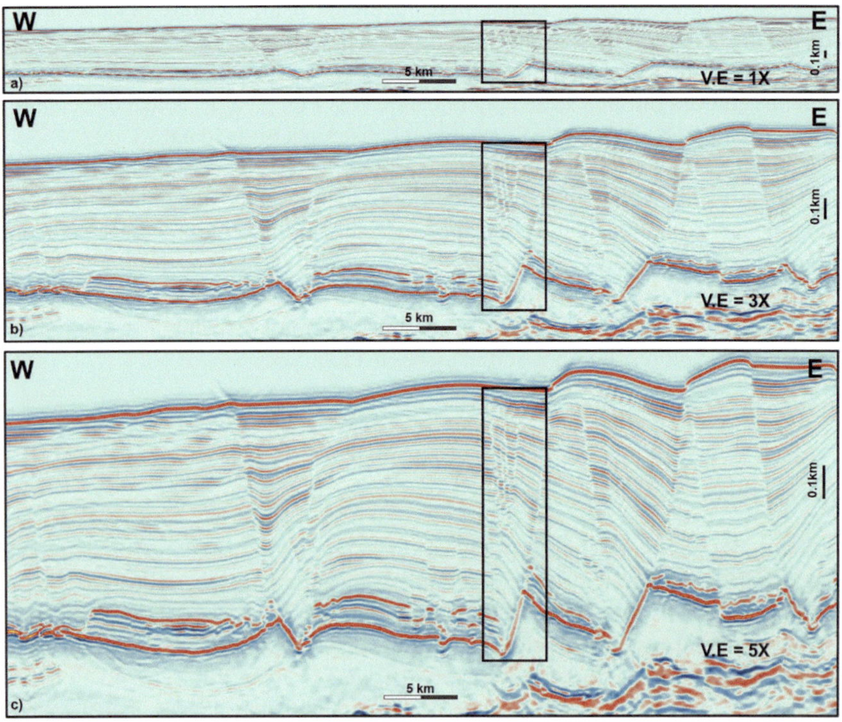

Fig. 2.10 Uninterpreted 2-D seismic profile showing the effects of vertical exaggeration (V.E.) on interpreting subsurface structures. **a** V.E. = 1 × gives a regional overview of fault geometry, stratigraphic architecture, and basin layout. **b** V.E. = 3 × increases visibility of subtle fault displacements and reflector terminations within the slope and basin areas. **c** V.E. = 5 × improves the resolution of steeply dipping features and rollover geometries within the growth fault system. Changes in V.E. influence the apparent steepness of faults, folds, and stratigraphic layers, thus affecting geological interpretation. Higher exaggeration (**c**) enhances vertical resolution and makes subtle structures more visible, while lower exaggeration (**a**) offers a more natural view of subsurface geometries. Black rectangles highlight regions of interest at higher vertical exaggeration, emphasising changes in structural expression

Table 2.4 Quick-reference guide for true dip, vertical exaggeration, and apparent dip

θ_{true}	VE	θ_{app} (approx.)
15°	3	40.6°
20°	3	49.4°
30°	3	60.0°
30°	5	73.3°
45°	2	63.4°

corrections were made to restore the true fault dip angles. The relationship between true dip, vertical exaggeration, and apparent dip is summarised in Table 2.4

Key Considerations

Although VE is a valuable visualisation tool, its application must balance visual clarity with geological accuracy. The optimal VE depends on the depth of interest, structural relief, and geological complexity. Overuse risks undermining interpretation, whereas careful application enhances structural mapping and the communication of results.

Worked Example: Apparent vs. True Dip Under Vertical Exaggeration

Geometry: Vertical exaggeration (VE) scales the vertical axis by a factor of VE while leaving the horizontal axis unchanged. As a result, the apparent dip (θapp) observed on a section with VE differs from the true dip (θtrue).

Forward problem (true → apparent):

$$\theta_{app} = \arctan(VE \times \tan \theta_{true})$$

Inverse problem (apparent → true):

$$\theta_{true} = \arctan((1/VE) \times \tan \theta_{app})$$

Example 1: Effect of moderate VE on a gentle dip.

Given $\theta_{true} = 30°$ and VE = 3,

$$\theta_{app} = \arctan\,(3 \times \tan 30°) = \arctan\,(1.732) \approx 60°$$

A gentle 30° fault appears twice as steep at VE = 3.

Example 2: Strong VE on a very gentle dip.

Given $\theta_{true} = 20°$ and VE = 5,

$$\theta_{app} = \arctan(5 \times \tan 20°) = \arctan(1.820) \approx 61.5°$$

A shallow 20° fault can appear ~ 62° at VE = 5, which may lead to a high risk of overestimating dip-slip components.

Example 3: Recovering the true dip from an exaggerated section.

If $\theta_{app} = 60°$ is observed at VE = 5,

$$\theta_{true} = \arctan((1/5) \times \tan 60°) = \arctan(0.346) \approx 19.1°$$

So, the true dip is only ~ 19°.

The effect increases rapidly for gentle dips (small tan θ) and is amplified by VE. Always report the VE used in figures, and for quantitative analysis, measure or calculate the dip on VE = 1 × sections.

Thickness and throw under VE.

- VE scales apparent thickness/throw on the section:
- $T_{app} = VE \times T_{true}$
- Horizontal separations (heave) are unchanged by VE.
- For kinematic or balanced cross-sections, calculate dips, throws, and bed thicknesses on s = 1 × plots or adjust them using the formulas provided.

Note: If a bed appears 100 m thick on a VE = 4 × section, the true thickness is 25 m. If a fault measures 200 m on VE = 5 ×, the true throw is 40 m.

At VE = 1 × , faults display their true dip, enabling precise kinematic analysis. At VE = 5 × , faults appear considerably steeper than they are. This exaggeration may cause misinterpretations of fault mechanics, such as overestimating extensional strain or misjudging fault dip-slip motion. Therefore, VE should be used cautiously for visualisation, but true-scale sections are vital for structural analysis.

2.5 Limitations of the Levantine Margin Seismic Data Set

A significant limitation of the Levantine Margin Seismic Data Set is the absence of well-log data for velocity calibration. Deepwater offshore drilling is extremely costly, restricting access to direct borehole control. Consequently, seismic correlation relied on arbitrary line interpretation, with lines drawn at random across the base map to establish reflector continuity. Validation involved cross-referencing intersecting inlines and crosslines, confirming consistency with recognised fault geometries, and incorporating seismic attributes such as coherence and curvature.

Interpretation challenges were especially severe in regions characterised by complex or chaotic reflections, often linked to evaporite movement and salt-related deformation. In these areas, multiple arbitrary lines were necessary to define reflector geometries. Even with a high-resolution 3D seismic dataset, reflector clarity and continuity were diminished, most likely due to low signal-to-noise ratios or subsalt imaging distortions. These limitations emphasise the necessity for additional well data and advanced imaging techniques to enhance structural and stratigraphic interpretations within salt-influenced basins.

2.6 Conclusion

1. Two key tectono-stratigraphic phases characterise the Levantine Margin:
   - the Messinian phase, immediately after evaporite deposition, and
   - the Plio–Quaternary phase, which documents the thin-skinned extensional response of the Levant Margin.
2. The tectonic evolution of the Levant Margin began with the breakup of Gondwana in the Palaeozoic and was heavily influenced by the Late Cretaceous collision between the Afro-Arabian and Eurasian plates. During the Late Miocene, northward movement of the Anatolian Plate activated the Levant Fracture System (LFS), which separated the Arabian and African plates and coincided with the opening of the Red Sea and the Gulf of Aden. The outer shelf and slope now exhibit thin-skinned extensional structures, including grabens, slumps, rotated fault blocks, listric growth faults, and salt rollers.
3. The 3D seismic imaging dataset covers an area of 5,372 km^2. The velocity model was initially built and then refined using wavelet-shift tomography (HyperTomo). Three key horizons, the seabed, the base of the Pliocene, and the base salt, anchor the depth conversion.
4. Interpreting the Levantine Margin dataset necessitated an integrated workflow.
   - establishing a georeferencing system,
   - analysing seismic resolution (amplitude, wavelength, acoustic impedance) to evaluate reflector clarity, and
   - Importing and visualising the 3D seismic volume in specialised geological software. These steps laid the foundation for reliable horizon picking, fault analysis, and structural interpretation.

2.7 Learning Objectives

After completing this chapter, readers will be able to:

1. Describe the acquisition, processing, and key features of high-resolution 3-D seismic data.
2. Explain how differences in acoustic impedance produce seismic reflections.
3. Evaluate the impact of scale, resolution, and vertical exaggeration on seismic interpretation.
4. Utilise velocity models to implement time-to-depth conversion concepts.
5. Critically assess the limitations of datasets in salt-affected and deep-water environments.

References

Bacon, M., Simm, R., & Redshaw, T. (2003). *3-D Seismic Interpretation*. Cambridge University Press.

Bacon, M., Simm, R., & Redshaw, T. (2007). *3-D Seismic Interpretation*. Cambridge University Press.

Bartov, Y. (1979). Israel—Geological Map: Survey of Israel, 1:500,000 scale.

Ben-Avraham, Z. (1978). The structure and tectonic setting of the Levant continental margin in the Eastern Mediterranean. *Tectonophysics, 46*(3–4), 313–331.

Ben-Avraham, Z., Ginzburg, A., Makris, J., & Eppelbaum, L. (2002). Crustal structure of the Levant Basin, eastern Mediterranean. *Tectonophysics, 346*(1–2), 23–43.

Bertoni, C., & Cartwright, J. A. (2006). Controls on the basinwide architecture of the late Miocene (Messinian) evaporite on the Levant margin (Eastern Mediterranean). *Sedimentary Geology, 188*, 93–114.

Beydoun, Z. R. (1999). Evolution and development of the Levant (Dead Sea Rift) Transform System: A historical-chronological review of a structural controversy. *Geological Society, London, Special Publications, 164*(1), 239–255.

Brown, A.R. (2012). *Interpretation of Three-Dimensional Seismic Data* (6th ed.). AAPG Memoir/SEG Investigations in Geophysics No. 9.

Cartwright, J. A., & Jackson, M. A. (2008). Initiation of gravitational collapse of an evaporite basin margin: The Messinian saline giant, Levant Basin, eastern Mediterranean. *Geological Society of America Bulletin, 120*, 399–413.

Clark, I. R., & Cartwright, J. A. (2009). Interactions between submarine channel systems and deformation in deepwater fold belts: Examples from the Levant Basin and the Eastern Mediterranean Sea. *Marine and Petroleum Geology, 26*, 1465–1482.

Clauzon, G., Suc, J.-P., Gautier, F., Berger, A., & Loutre, M.-F. (1996). Alternate interpretation of the Messinian salinity crisis: Controversy resolved? *Geology, 24*(4), 363.

Donnellan, A., Arrowsmith, R., & DeLong, S. (2017). Spatio-temporal mapping of plate boundary faults in California using geodetic imaging. *Geosciences, 7*(1), 15. https://doi.org/10.3390/geosciences7010015

Druckman, Y., Buchbinder, B., Martinotti, G. M., Tov, R. S., & Aharon, P. (1995). The buried Afiq Canyon (eastern Mediterranean, Israel): A case study of a Tertiary submarine canyon exposed in Late Messinian times. *Marine Geology, 123*(3–4), 167–185.

Elfassi, Y., Gvirtzman, Z., Katz, O., & Aharonov, E. (2019). Chronology of post-Messinian faulting along the Levant continental margin and its implications for salt tectonics. *Marine and Petroleum Geology, 109*, 574–588.

Freund, R. (1975). The Triassic-Jurassic structure of Israel and its relation to the origin of the eastern Mediterranean. *Geological Survey of Israel.*

Garcia-Castellanos, D., Micallef, A., Estrada, F., Camerlenghie, A., Ercillad, G., Periáñez, R., & Abril, J. M. (2020). The Zanclean mega-flood of the Mediterranean—Searching for independent evidence. *Earth-Science Reviews, 201*, 1–14.

Gardosh, M., Druckman, Y., Buchbinder, B., & Rybakov, M. (2006). The Levant Basin offshore Israel: stratigraphy, structure, tectonic evolution and implications for hydrocarbon exploration. *Geophysical Institute of Israel*, 1–119.

Garfunkel, Z. (1981). The internal structure of the Dead Sea leaky transform (rift) in relation to plate kinematics. *Tectonophysics, 80*, 81–108.

Garfunkel, Z. (1984). Large-scale submarine rotational slumps and growth faults in the eastern Mediterranean. *Marine Geology, 55*, 305–324.

Garfunkel, Z. (1998). Constraints on the origin and history of the Eastern Mediterranean basin. *Tectonophysics, 298*, 5–35.

PGS Geophysical 2013a. Acquisition Report of MC3D-LEB 2013 Seismic Reflection Data. Unpublished. Version 1, pp. 6–9.

PGS Geophysical 2013b. Processing Report of MC3D-LEB 2013 Seismic Reflection Data. Unpublished. Project no. 3212, pp. 7–12.

Ghalayini, R., Nader, F. H., Bou Daher, S., Hawie, N., & Chbat, W. E. (2018). Petroleum systems of Lebanon: An update and review. *Journal of Petroleum Geology, 41*(2), 189–214.

Gradmann, S., Hübscher, C., Ben-Avraham, Z., Gajewski, D., & Netzeband, G. (2005). Salt tectonics off northern Israel. *Marine and Petroleum Geology, 22*, 597–611.

Gradstein, F.M., Ogg, J.G., Schmitz, M. & Ogg, G. (eds.) (2012). The geologic time scale 2012. *Elsevier*.

Gvirtzman, Z., Reshef, M., Buch-Leviatan, O., & Ben-Avraham, Z. (2013). Intense salt deformation in the Levant Basin in the middle of the Messinian salinity crisis. *Earth Planet. Science Letters, 379*, 108–119.

Gvirtzman, Z., Manzi, V., Calvo, R., Gavrieli, I., Gennari, R., Lugli, S., Reghizzi, M., & Roveri, M. (2017). Intra-Messinian truncation surface in the Levant Basin explained by subaqueous dissolution. *Geology, 45*, 915–918.

Hatton, W., McIntosh, K., & Edwards, J. (2003). The application of vertical exaggeration in seismic interpretation. In *SEG Technical Program Expanded Abstracts 2003* (pp. 979–982). Society of Exploration Geophysicists.

Hempton, M. R. (1987). Constraints on Arabian plate motion and extensional history of the Red Sea. *Tectonics, 6*(6), 687–705.

Hsü, K., Ryan, W., & Cita, M. (1973). Late Miocene desiccation of the Mediterranean. *Nature, 242*, 240–244.

Hubral, P., & Tygel, M. (1997). Depth imaging. *Geophysical Journal International, 129*(1), 1–15.

Müller, T. M., & Müller, B. (2020). Recognising the limits of seismic interpretation. *The Leading Edge, 39*(8), 620–628.

Nwosu, O. (2013). *The Compartmentalization of a Deepwater Fold and Thrust Belt in the Levant*. (PhD thesis). Cardiff University: UK.

Onajite, E. J. (2014). *Seismic Data Analysis Techniques in Hydrocarbon Exploration* (pp. 198–202). Elsevier.

Roberts, G., & Peace, D. (2007). Hydrocarbon plays and prospectivity of the Levantine Basin, Levantine Margin, and Syria from modern seismic data. *Geo-Arabia, 12*(3), 99–124.

Ryan, W. B., & Cita, M. B. (1978). The nature and distribution of Messinian erosional surfaces—indicators of a several-kilometre-deep Mediterranean in the Miocene. *Marine Geology, 27*, 193–230.

Ryan, W., Hsü, K., Nesteroff, W., Pautot, G., Wezel, F., Lort, J., Cita, M., Mayne, W., & Stradner, H. (1973). *Initial Reports of the Deep-Sea Drilling Project* (Vol. 13). US Washington Printing: Washington.

Sheriff, R. E., & Geldart, L. P. (1995). *Exploration Seismology* (2nd ed.). Cambridge University Press.

Skiple, C., Anderson, E., & Fürstenau, J. (2012). Seismic interpretation and attribute analysis of the Herodotus, Levantine Basin, offshore Cyprus, and Lebanon. *Petroleum Geoscience, 18*, 433–442.

Stewart, S. A. (2011). Vertical exaggeration of reflection seismic data in geoscience publications 2006–2010. *Marine and Petroleum Geology, 28*(5), 959–965.

Stewart, S. A. (2012). Interpretation validation on vertically exaggerated reflection seismic sections. *Journal of Structural Geology, 41*, 38–46.

Stone, D. S. (1991). Analysis of scale exaggeration on seismic profiles. *AAPG Bulletin, 75*(7), 1161–1177.

Tibor, G., & Ben-Avraham, Z. (2005). Late Tertiary paleodepth reconstruction of the Levant margin off Israel. *Marine Geology, 221*, 331–347.

Walley, C. D. (1998). Some outstanding issues in the geology of Lebanon and their importance in the tectonic evolution of the Levantine region. *Tectonophysics, 298*, 37–62.

SEG Wiki (2024). Seismic resolution. https://wiki.seg.org/wiki/Seismic_resolution

Wu, G., & Wu, S. (2017). Depth-to-horizon calculation and its applications in seismic interpretation. *Interpretation*, *5*(4), SMi41-SMi56.

Yilmaz, O. (2001). Seismic data analysis—processing, inversion, and interpretation of seismic data. *SEG, Tulsa.* https://doi.org/10.1190/1.9781560801580

Chapter 3
Seismic Data Mapping

Abstract This chapter guides you through a complete, reproducible workflow for interpreting high-resolution 3-D seismic reflection data from the Levantine Margin. It proceeds step by step from seismic visualisation and scale optimisation to horizon picking, fault identification, and the generation of time- and depth-structure maps. You learn how reflector character, amplitude, continuity, and acoustic impedance contrasts govern horizon confidence, and how choices in display scale, vertical exaggeration, and velocity modelling directly influence geological interpretation. Worked examples demonstrate practical strategies for mapping continuous and discontinuous reflectors, minimising mis-ties, and addressing imaging challenges in salt-affected, thin-skinned extensional systems. Fault-mapping criteria are clearly defined and applied to build a coherent three-dimensional structural framework. Throughout the chapter, emphasis is placed on quality control, uncertainty awareness, and iterative refinement. Interactive prompts encourage you to question interpretation choices, test alternative scenarios, and link seismic observations to rift architecture and fault kinematics. By the end of the chapter, you will be able to construct dependable 3-D seismic frameworks and transfer these skills to other rifted and salt-influenced continental margins.

Keywords Seismic interpretation · Time–depth conversion · Acoustic impedance · Thin-skinned extension · Salt tectonics · Levant margin

3.1 Introduction

This chapter explores methods for analysing high-resolution 3-D seismic data from the Plio–Quaternary strata of the Levantine margin in the eastern Mediterranean. My goal is clear: to identify and map key seismic reflections (stratal horizons) and the faults that displace them. Once recognised, these horizons are depicted as three-dimensional surfaces, contoured for depth, and visualised through time-structure and depth-structure maps. Additionally, fault-pattern maps are created to demonstrate the geometry and distribution of deformation.

K. Chiadikobi, *Advances in Horizon Mapping and Fault Interpretation*,
https://doi.org/10.1007/978-3-032-23917-4_3

The workflows outlined here follow established seismic interpretation practices described by leading authors (Aki & Richards, 2002; Bacon et al., 2003; Brown, 2012; Onajite, 2014). If you are already familiar with seismic interpretation, you might recognise several of these steps. If not, this chapter will guide you through them step by step.

One of the ongoing challenges in 3-D seismic interpretation is identifying individual reflectors that correspond to separate stratal surfaces. As you will observe, this task becomes especially difficult in regions with poor seismic imaging or where geological structures disrupt continuity, such as faults or steep monoclines (Bacon et al., 2003; Brown, 2012). These complexities increase the risk of mis-ties, which, in turn, can undermine the accuracy of horizon maps.

How can we minimise these risks? In the following sections, I outline the workflows designed to reduce uncertainty in horizon mapping. You will also learn about the approaches used for fault mapping, which together provide the foundation for reconstructing the rift architecture of the Levantine margin.

3.2 Seismic Horizon Mapping

As introduced in the previous chapter, it is covered by a high-resolution 3D seismic dataset. In this chapter, I apply established seismic interpretation methods to this volume, focusing on two main tasks: (1) mapping stratal reflectors and associated faults, and (2) constructing time-contoured and fault distribution maps. These methods follow the principles described by Brown (2012) and Onajite (2014).

Horizon mapping aims to identify and trace laterally consistent reflectors that indicate geological structures and stratigraphic boundaries. A practical question arises: where should we start? In most cases, the most dependable starting point is the seabed horizon at the top of the seismic section. At this level, reflectors are usually well defined and continuous. From there, the interpretation moves downward, step by step, into intervals where reflectors become less clear, and interpretation demands more judgement.

3.2.1 Step-By-Step Horizon Mapping Workflow

Interpretation of reflectors within the 3-D seismic volume followed a structured workflow using specialised geological and geophysical software. Each step was designed to enhance visibility, reduce uncertainty, and ensure consistency, and was integrated into three-dimensional stratal frameworks (see Fig. 3.1). The procedures are summarised below:

Fig. 3.1 Horizon mapping workflow in 3D seismic interpretation

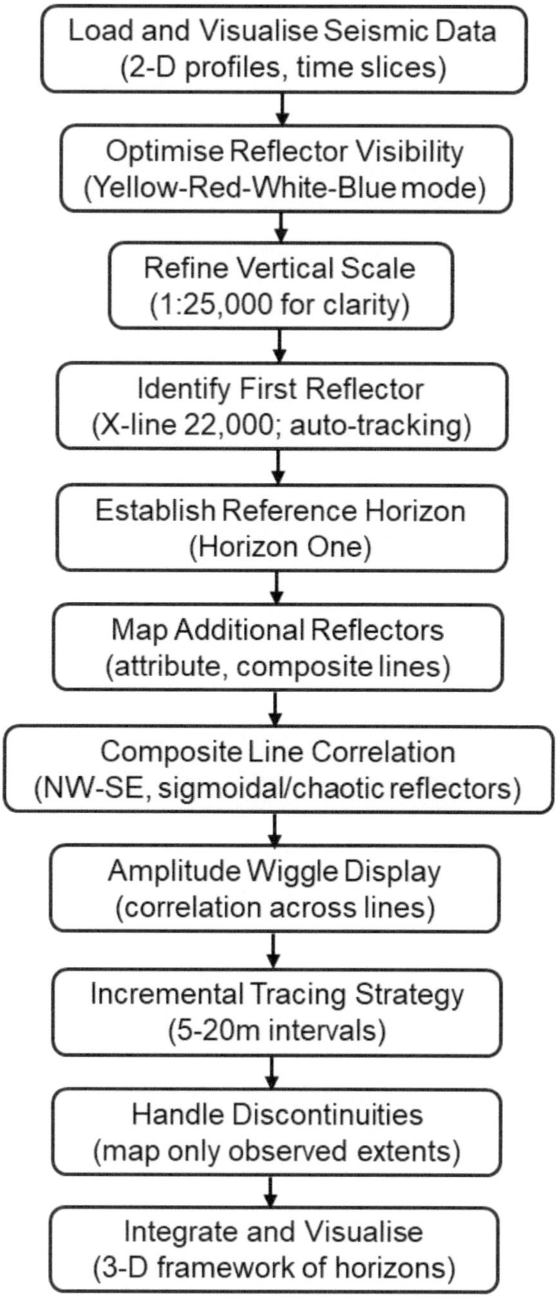

1. Initial Examination of the Seismic Volume

   The 3-D seismic dataset was imported into geological software. Seismic profiles were visualised, including vertical inlines, crosslines, and horizontal time slices. This provided a foundation for identifying suitable starting points for reflector mapping.
2. Optimising Reflector Visibility

   The seismic display was adjusted to the Yellow–Red–White–Blue Turquoise (compressed) colour scheme. This mode improved amplitude contrasts and enhanced reflector transparency, making subtle stratigraphic features easier to recognise (see Fig. 3.2a and b).
3. Adjusting Seismic Scale

   The vertical scale was standardised to 1:25,000. At this resolution, reflector geometries appeared clearer, aiding in confident interpretation of reflector terminations and thickness variations (see Fig. 3.3).

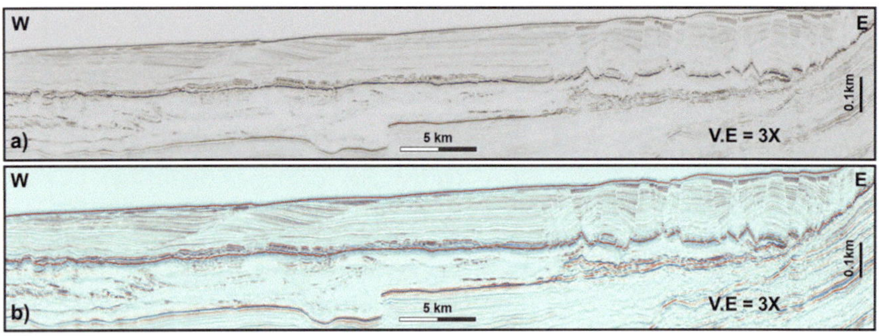

Fig. 3.2 Uninterpreted seismic profiles with colour images: **a** Seismic (default); **b** Yellow, red, white, blue, turquoise (compressed)

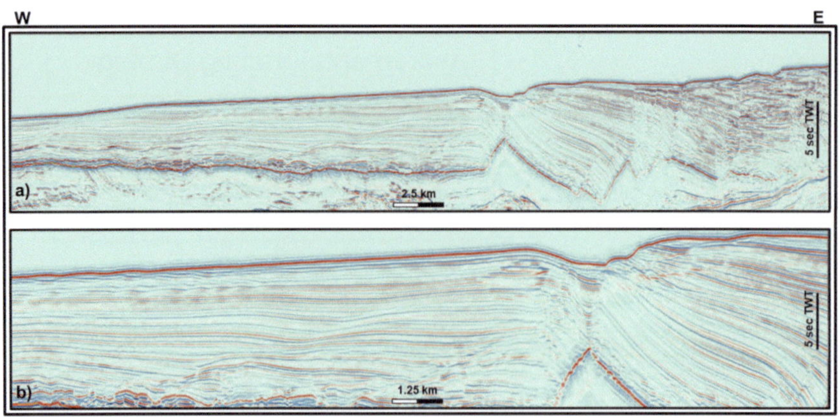

Fig. 3.3 Uninterpreted seismic profiles displaying different scales of the seismic section, **a** 1:50,000, **b** 1:25,000

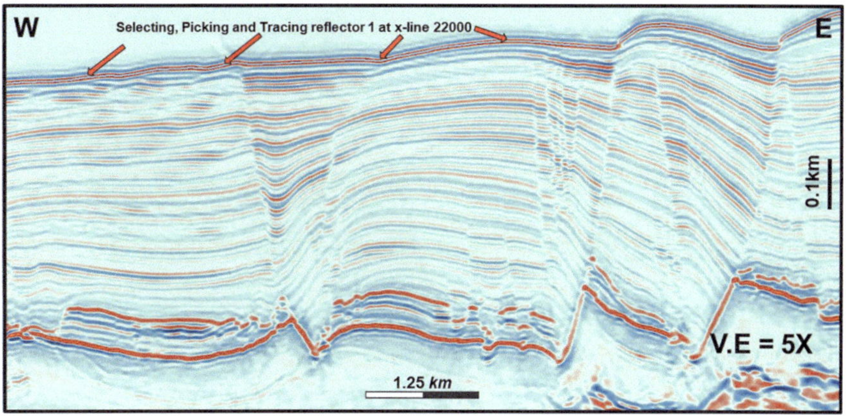

Fig. 3.4 Selecting, picking, and tracing reflector 1 (Seabed) on the seismic section

4. Selection of the First Reflector
   The first key reflector was identified at Line 22,000 and moved gradually from west to east. It showed high amplitude, strong continuity, and favourable impedance contrasts. The seeded 3-D auto-tracking tool was then used to trace it through the seismic volume (see Fig. 3.4).
5. Establishing a Reference Horizon
   This initial reflector was named Horizon One. As a reference horizon, it acted as a structural anchor for mapping subsequent reflectors and served as a baseline for cross-sectional correlation.
6. Mapping Subsequent Reflectors
   In the absence of well-log calibration, reflector identification depended on seismic attributes, specifically amplitude, continuity, resolution, and impedance contrasts. Arbitrary (composite) lines were drawn across the dataset to ensure consistency in reflector correlation.
7. Use of Composite Lines for Complex Geometries
   Composite lines were especially useful in areas with irregular seismic signatures. These lines linked multiple regions across the base map (see Fig. 3.5) and allowed for more confident tracking of reflectors with discontinuous, sigmoidal, or chaotic patterns (Bacon et al., 2003; Brown, 2012; Onajite, 2014).
8. Selection of the Second Reflector
   The second reflector was identified along arbitrary (composite) lines, oriented NW-SE across the seismic section. It displayed moderate to high amplitudes and favourable impedance contrasts (see Fig. 3.6). Using the 2-D auto-tracking horizon tool, it was carefully traced along the composite lines to establish continuity.
9. Amplitude Wiggle Display for Correlation

Fig. 3.5 Creating arbitrary (composite) lines by drawing a random line connecting different areas of the base map

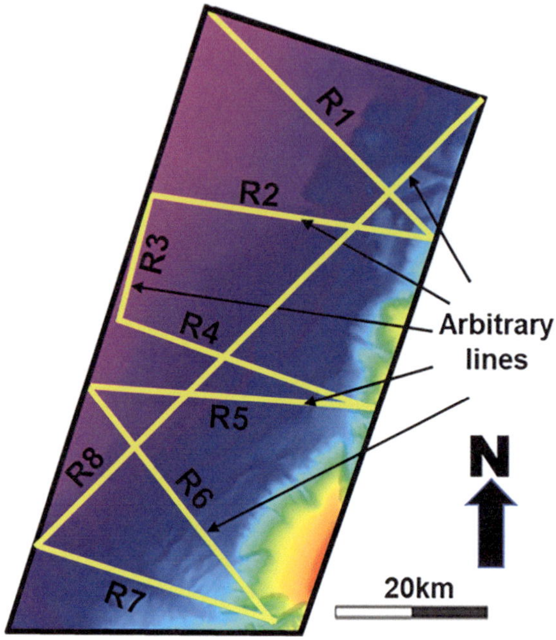

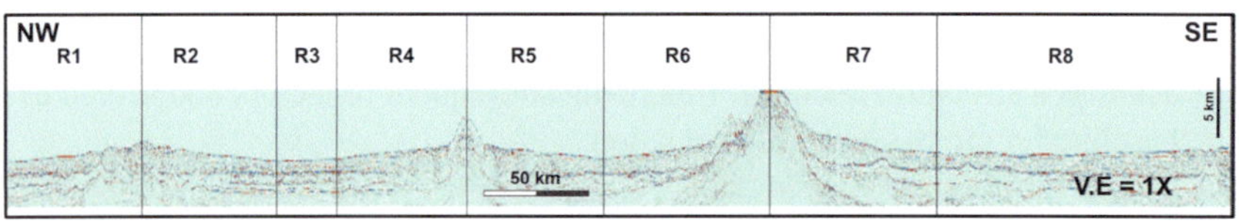

Fig. 3.6 Selecting other reflectors on arbitrary (composite) lines in 2-D profiles

To improve reflector identification, "Amplitude Wiggles" were created along the seismic sections. These visualisations emphasised amplitude variations, which were especially helpful for correlating equivalent reflectors across multiple seismic lines. This step enhanced both the accuracy and consistency of reflector interpretation (see Fig. 3.7).

10. Incremental Reflector Tracing

    A tracing strategy was applied to a second reflector along Line 22,000, progressing from west to east. Reflectors were tracked at intervals of 5–20 m (e.g., 12.5 m and 50 m). This incremental mapping helped establish robust correlations among stratal reflectors. Multiple interpretation tools, including seeded 2-D auto-tracking horizons, were used to ensure precision.

11. Handling Discontinuous Reflectors

    Not all reflectors could be mapped as continuous horizons. Forcing continuity across disrupted or poorly imaged zones risked introducing misalignments, especially when checked against crosslines or arbitrary lines. To prevent

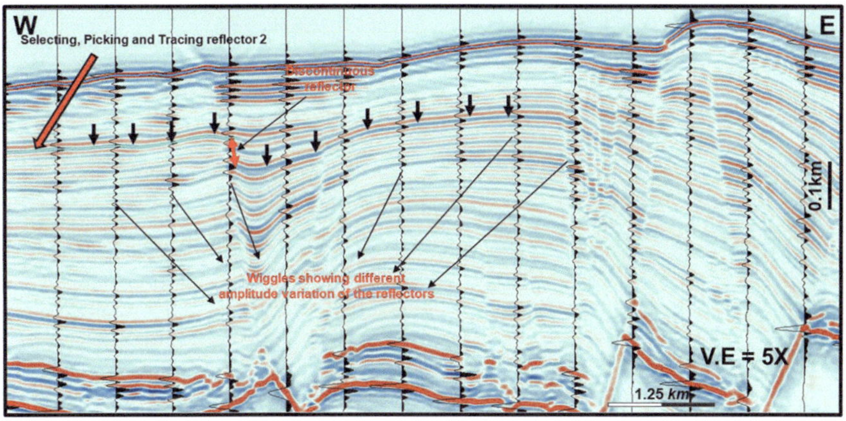

Fig. 3.7 Amplitude wiggle variations and intensities on the seismic section

this, discontinuous reflectors were mapped only within their observed extents. Gaps, breaks, or abrupt terminations were left unmapped to maintain geological integrity (see Fig. 3.8).

12. Application to Additional Reflectors

    The same methodology, incremental tracing, amplitude validation, and careful handling of discontinuities, was applied to subsequent reflectors across the seismic section. This ensured a consistent workflow and reproducible results.

13. Integration and Visualisation

    Finally, all mapped horizons were integrated into a comprehensive 3-D framework (see Fig. 3.9). This cumulative visualisation provided a holistic view of reflector geometries, offering structural and stratigraphic insights into the Levantine margin.

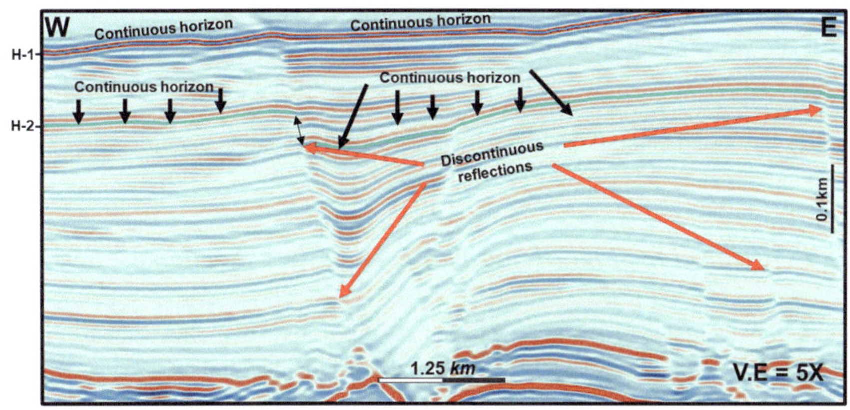

Fig. 3.8 Visualisation of horizon two (2) showing continuous and discontinuous horizons

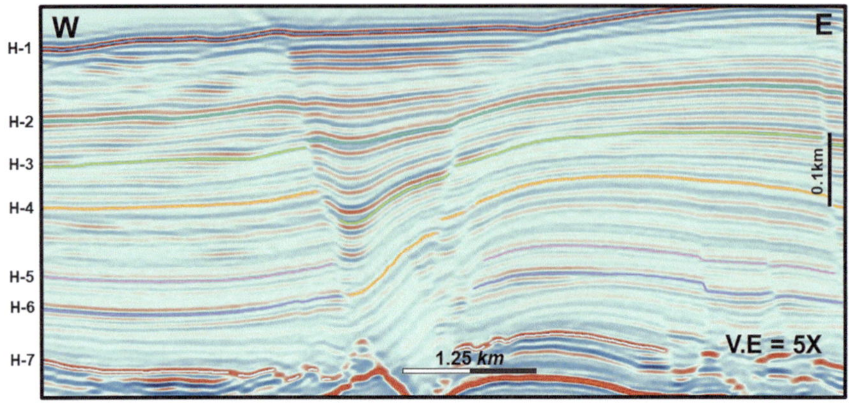

Fig. 3.9 Visualisation of eight horizons mapped on **a** the seismic section

This workflow shows that careful horizon mapping, using a combination of auto-tracking, amplitude validation, incremental spacing, and manual judgment, reduces uncertainty and prevents mis-ties. By respecting reflector discontinuities and cross-validating across multiple lines, interpreters produce reliable 3-D frameworks that support subsequent fault mapping and structural analysis.

3.3 Fault Mapping

Fault mapping was conducted on seismic profiles to outline the basin's structural framework and to position it within its tectono-stratigraphic context (Bacon et al., 2003; Brown, 2012; Onajite, 2014; Yilmaz, 2001). Structural styles, as described by Lowell (1985), are typical associations of faults and folds formed by shared geodynamic processes. The focus was on identifying essential fault geometries, including listric, planar, and block-bounding faults, within the Levantine margin.

3.3.1 Criteria for Identifying Faults in Seismic Sections

Faults were identified based on established seismic criteria (e.g., Bacon et al., 2003; Onajite, 2014). Four key diagnostic features guided the interpretation:

1. Correlatable Offsets of Stratal Reflectors
   Faults were identified where continuous seismic reflectors ended suddenly and then reappeared in shifted positions across fault planes. In 3D seismic volumes, vertically stacked reflector terminations retained the sense of displacement (normal versus reverse faulting; see Fig. 3.10a-i to iv).
2. Abrupt Changes in Dip and Seismic Character
   Faults were identified where horizon terminations created arcuate reflections that intersected with adjacent reflectors, thus modifying dip directions and seismic characteristics. For instance, on a 2-D seismic profile, a reflector might dip landward with mostly continuous westerly patterns, then switch to basinward dips with sigmoidal or discontinuous patterns (see Fig. 3.10b-i, ii).
3. Stacked Narrow Monoclines Through Multilayer Reflectors
   Some faults appeared as narrow, convex-upward monoclines cutting across several reflector sets. These features could either represent genuine structural geometries or artefacts caused by limited seismic resolution. Vertical exaggeration ($\times 3$) was applied to emphasise subtle offsets and reflector dips that were otherwise indistinct (see Fig. 3.10c-i, ii).
4. Stacked Narrow Zones of Dimmed Amplitude
   Vertically aligned zones of reduced amplitude served as an additional diagnostic criterion. In these cases, reflector continuity was weakened across multiple layers, often indicating localised deformation zones (see Fig. 3.10d-i, ii).

3.3.2 Fault Mapping Workflow

A workflow was used to identify faults throughout the 3D seismic volume (see Fig. 3.11).

1. Review Horizon Interpretations
   Horizon maps were initially re-evaluated to identify potential regions for fault initiation.
2. Optimise Seismic Display
   The vertical display scale was adjusted to 1:30,000 to improve the visibility of subtle discontinuities and reflector terminations.
3. Select Key Profile for Initial Fault Mapping
   Line 22,000 was selected due to its distribution of discontinuous reflectors (see Fig. 3.12a). Faults were traced progressively from shallow to deeper sections, using reflector offsets as reference points. Tracing was carried out at 2 km intervals to maintain consistency.
4. Extend Tracing Across the Seismic Grid
   The same method was applied across different inlines and crosslines to ensure structural consistency throughout the seismic volume.

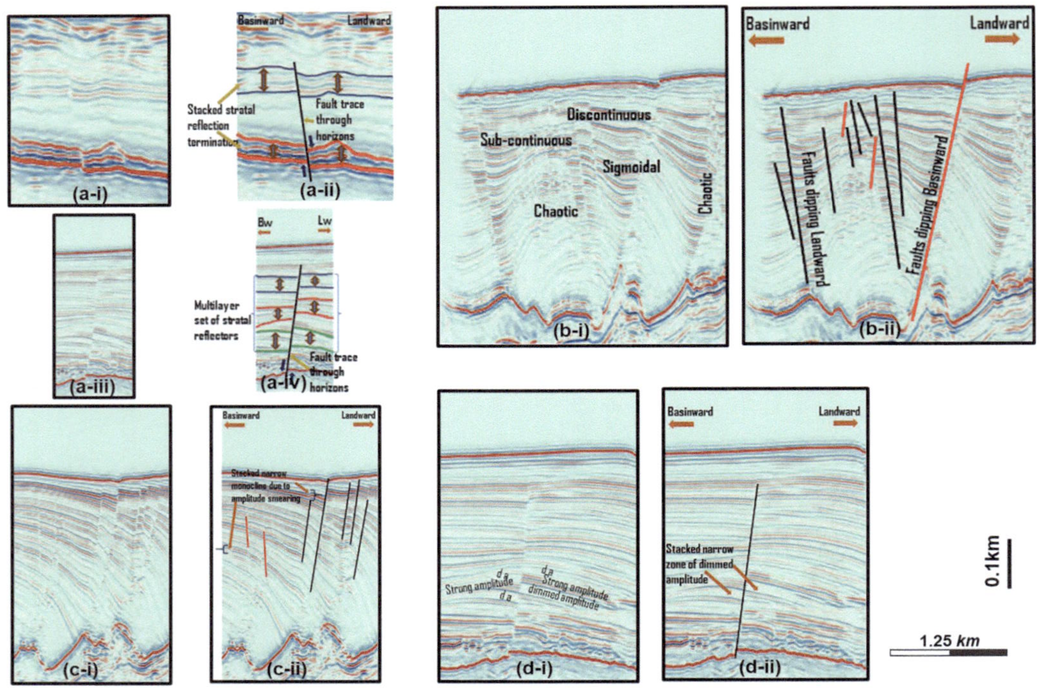

Fig. 3.10 Criteria for Identifying Faults in Seismic Sections. (**a**-i) to (**a**-iv) show various seismic features, including stacked stratal reflection terminations and fault traces crossing multiple stratigraphic layers. (**b**-i) and (**b**-ii) illustrate sudden changes in dip and seismic patterns. The figure also displays variability in seismic facies, such as discontinuous, sub-continuous, sigmoidal, and chaotic reflections. (**c**-i) and (**c**-ii) depict a narrow monocline structure affected by amplitude smearing and faulting. (**d**-i) and (**d**-ii) present amplitude variations, including stacked narrow zones of strong and dimmed amplitude

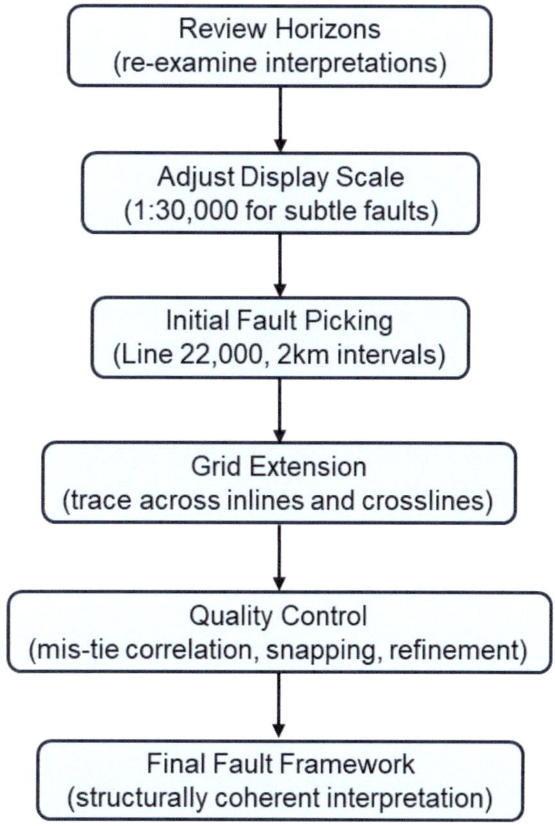

Fig. 3.11 Fault mapping workflow in 3D seismic interpretation

3.3.3 Quality Control of Fault Interpretation

Following initial horizon and fault mapping, quality-control steps were undertaken:

- Correction of Horizon Mis-ties: Localised mismatches, often observed as "spikes," were corrected to restore continuity (see Fig. 3.12b).
- Snapping Horizons to Fault Planes: Horizons were reconnected to adjacent fault planes to ensure structural coherence (see Fig. 3.13a and b).
- Refinement of Fault-Horizon Relationships: Fault throw, continuity, and reflector alignment were re-verified to enhance the accuracy of the final interpretation.

Fault interpretation relies on reliable horizon mapping. By combining reflector offsets, dip changes, monocline geometries, and amplitude anomalies, and

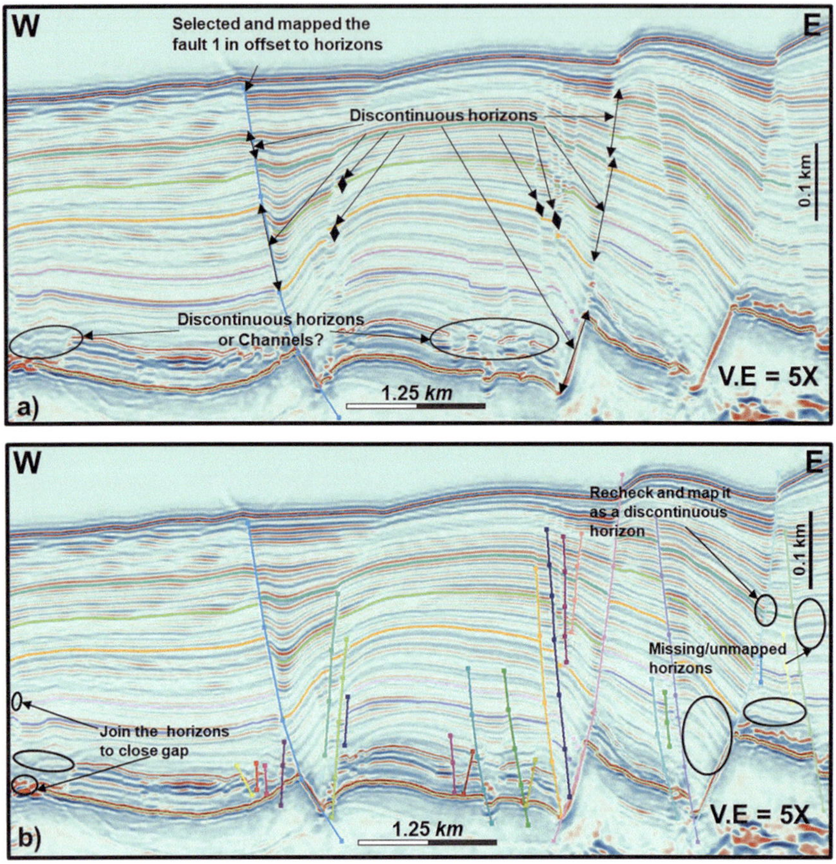

Fig. 3.12 **a** Stages in mapping faults on the seismic section and **b** Visualisation of the quality check process for horizons on the seismic section

by applying rigorous QC, interpreters minimise artefacts and build a structurally coherent framework. These mapped faults define the rift architecture and underpin displacement analysis across the Levantine margin.

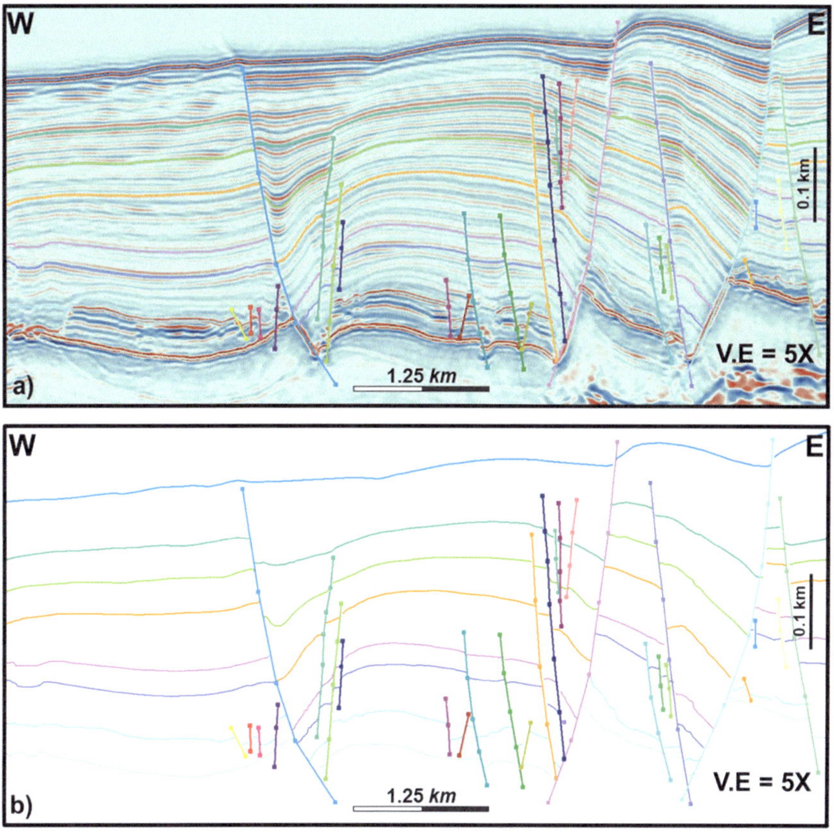

Fig. 3.13 Visualisation of the completed mapped horizons and faults, **a** with the seismic section, **b** without the seismic

3.4 Seismic Surface (Time-Structure) Mapping

Once horizon mapping was completed, seismic surfaces were generated for each interpreted horizon. These surfaces provided a framework for validating interpretations and enabled sequential analysis of individual mapped horizons. The resulting surfaces were contoured in two-way travel time (TWT), producing time-structure maps. These time-structure maps were later converted into depth-structure maps using the velocity model discussed in Chap. 2. Subsection 3.4.1 describes the workflow for seismic surface mapping (see Fig. 3.14).

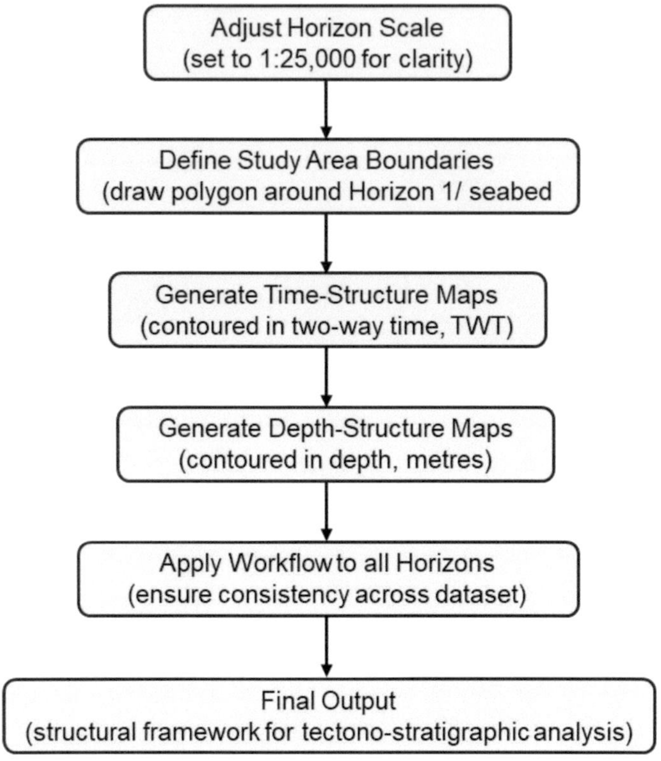

Fig. 3.14 Seismic Surface Mapping Workflow

3.4.1 Workflow for Seismic Surface Mapping

1. Adjust Horizon Scale for Mapping
   To improve visibility and clarity in the 2D map view, the display scale for each horizon was set to 1:25,000. This ensured that mapped reflectors appeared clear and prominent during interpretation (see Fig. 3.15a).
2. Define Boundaries
   A polygon was drawn around Horizon 1 (the seabed) to isolate the mapped horizons from the wider survey dataset. This step defined the boundary and prevented unwanted artefacts from affecting surface generation (see Fig. 3.15a).
3. Generate Time-Structure Maps
   Using the polygon constraints, surface maps were created for each horizon. These were contoured in TWT to visualise the structural geometry of each surface (see Fig. 3.15b).
4. Generate Depth-Structure Maps
   After generating time-structure surfaces, each horizon was depth-converted using the velocity model to produce depth-structure maps (see Fig. 3.15c).

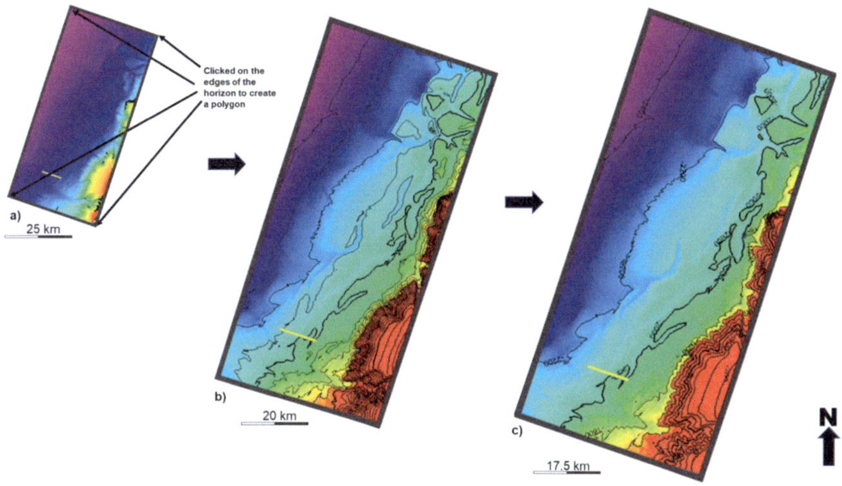

Fig. 3.15 **a** Stages in creating the polygon of the seabed, **b** Time-structure map, and **c** Depth-structure map

5. Apply Workflow Across All Horizons

   The same procedures were applied to all additional mapped horizons, ensuring consistency in scale, polygon definition, and surface generation throughout the seismic volume.

Time-structure maps link horizon interpretation with structural analysis. By depicting horizons as contoured surfaces, interpreters can efficiently identify fault-bounded blocks, tilted strata, and large-scale basin architectures. This method also offers a dependable basis for integrating fault interpretations and reconstructing the rift's tectono-stratigraphic evolution.

3.5 Producing Fault Maps

A comprehensive quality control (QC) review of surface maps and fault interpretations was carried out across the 3-D seismic volume to ensure structural consistency and accuracy. The interpreted faults were then imported into a structural modelling environment, where they formed the basis for a fault framework model. Subsection 3.5.1 details the workflow for producing fault maps (see Fig. 3.16).

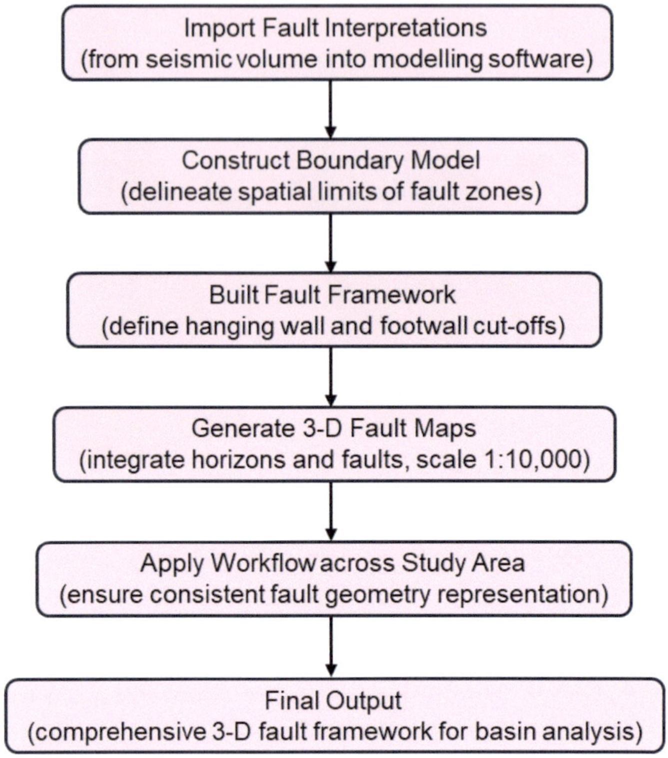

Fig. 3.16 Producing fault maps workflow

3.5.1 Workflows for Producing Fault Maps

1. Import Fault Interpretations

   Faults interpreted directly from the seismic volume were imported into a structural modelling platform. These interpretations defined the framework's core structural elements, including fault planes, bounding surfaces, and associated horizons.
2. Construct Boundary Model

   A boundary model was created to define the spatial extent of the faulted zones, offering a structural envelope for subsequent modelling (see Fig. 3.17a).
3. Build Fault Framework

   Using the boundary constraints, cutoff geometries for the hanging wall and footwall were defined from previously mapped horizons (see Fig. 3.17b–d).
4. Generate 3-D Fault Maps

   A detailed three-dimensional fault map was constructed by integrating horizon and fault geometries. To improve clarity and interpretation, the display scale was set to 1:10,000 (see Fig. 3.17e).

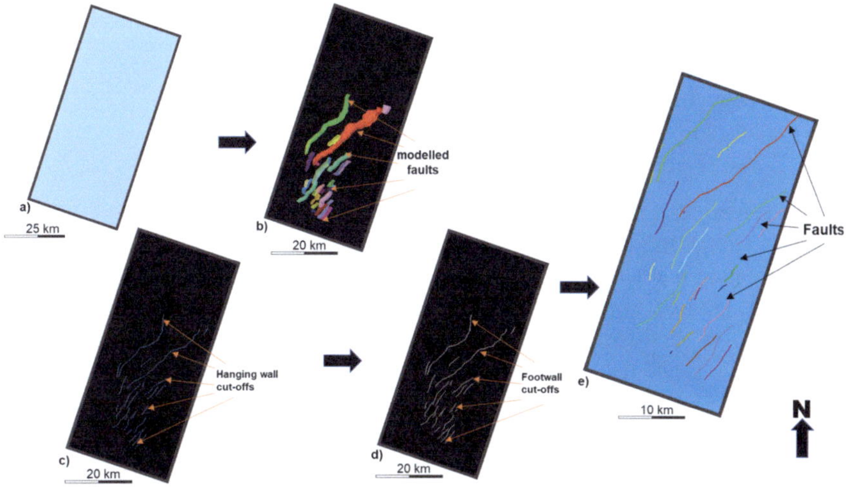

Fig. 3.17 Workflow for producing fault maps. **a** Blank framework for model initialisation. **b** Modelled fault surfaces extracted from seismic interpretation. **c** and **d** Identification of key structural elements, including hanging-wall and footwall cut-offs, essential for structural restoration. **e** Final interpreted fault network, with multiple fault traces extracted for further geomechanical analysis

5. Apply Workflow Across the Selected Area
   The same modelling procedure was repeated throughout the seismic volume, ensuring a consistent representation of fault geometry and structural interpretation.

This workflow integrated mapped faults into a 3D model by combining seismic interpretations with structural modelling, enabling detailed visualisation of fault geometries, terminations, and boundaries. The resulting 3D fault maps provided a robust framework for analysing rift architecture, fault kinematics, and basin evolution across the Levantine margin.

3.6 Conclusion

1. High-resolution 3-D seismic data from the Plio–Quaternary to Messinian strata of the Levantine margin form the basis of this chapter.
2. Reflectors selected for mapping were chosen for their high amplitudes and regional continuity, thereby reducing interpretation uncertainties and enhancing reflector-to-reflector correlations.

3. Faults were recognised on seismic profiles using four principal diagnostic criteria:
   - Correlatable offsets of stratal reflectors,
   - Abrupt changes in dip direction and seismic character,
   - Stacked monoclines cutting across multilayer reflectors, and
   - Zones of dimmed amplitude through multiple layers.

   These features confirmed the presence of steeply dipping faults that substantially offset stratigraphic reflections.
4. Partitioning the 3-D seismic volume into horizontal slices allowed fault traces to be visualised in plan view. This clarified fault network geometry, enhanced imaging of time-structure horizons, and facilitated quick assessment of fault intersections and depth-dependent changes on vertical sections.
5. Fault maps were produced for each mapped horizon. These maps displayed lateral variations in structural styles (such as listric, planar, and conjugate faults) and fault interactions, including relay ramps, cross-cutting, intersecting, and splay geometries. These observations offer a detailed understanding of rift architecture and fault kinematics within the Levantine margin.

Learning Objectives

After completing this chapter, readers will be able to:

1. Apply systematic workflows for seismic visualisation and scale optimisation.
2. Identify reliable starting horizons and justify horizon selection strategies.
3. Construct time-structure and depth-structure maps from interpreted horizons.
4. Integrate horizon and fault mapping into a coherent 3-D structural framework.
5. Implement quality-control procedures to minimise interpretation artefacts.

References

Aki, K., & Richards, P. (2002). *Quantitative seismology*. MIT Press.

Bacon, M., Simm, R., & Redshaw, T. (2003). *3-D seismic interpretation*. Cambridge University Press, First edition, New York.

Brown, A.R. 2012. Interpretation of three-dimensional seismic data. Sixth ed. (AAPG Memoir/SEG Investigations in Geophysics No. 9).

Lowell, J. D. (1985). *Structural styles in petroleum geology*. Oil and Gas Consultants International Inc.

Onajite, E. J. (2014). *Seismic data analysis techniques in hydrocarbon exploration* (pp. 198–202). Elsevier.

Yilmaz, O. (2001). Seismic data analysis—processing, inversion, and interpretation of seismic data. *SEG, Tulsa*. https://doi.org/10.1190/1.9781560801580

Chapter 4
Understanding Regional 2D Profiles Workflows

Abstract This chapter shows how regional structural and stratigraphic understanding can be effectively developed from regional two-dimensional (2D) seismic profiles, even when full three-dimensional (3D) coverage is limited. Using selected regional profiles taken from a 3D seismic volume of the Levantine Margin, the chapter reconstructs basin-scale cross-sections and fault arrays using traditional 2D interpretation methods for rifted continental margins. The workflow combines analogue interpretation with digital refinement to create a consistent stratigraphic–structural framework for the Plio–Quaternary and Messinian successions of the Levantine Margin. Four seismic–stratigraphic units (SP1–SP4) are identified based on reflector terminations, amplitude characteristics, internal geometry, and regional bounding horizons, revealing vertical and lateral variations across the basin. These units highlight differences in depositional style, fault influence, and syn-kinematic deformation between the eastern and western margins. Throughout the chapter, readers are encouraged to critically evaluate what structural complexity can be discerned from 2D data alone, how interpretation confidence changes with scale, and the limitations of 2D analysis. By the end, the chapter offers both a practical methodology and a reflective framework for interpreting rifted margins using regional 2D seismic datasets.

Keywords Seismic profiles · Stratigraphic–structural framework · Seismic–stratigraphic units · Messinian evaporites · Analogue-to-digital workflow · Facies analysis

4.1 Introduction

Structural mapping along rifted continental margins often depends on limited datasets. In many instances, researchers must rely on long 2D seismic profiles instead of full 3D volumes. This chapter explores how this limitation influences the quality and depth of structural interpretation. To examine this, I extracted several long 2D seismic profiles from the available 3D seismic volume of the Levantine Margin and used them to reconstruct regional cross-sections and fault-array geometries. You can

K. Chiadikobi, *Advances in Horizon Mapping and Fault Interpretation*,
https://doi.org/10.1007/978-3-032-23917-4_4

think of this as a deliberate "scaling down" of perspective. Rather than mapping every fault interaction in 3D, I use the approaches traditionally applied to interpret other rifted margins with only 2D coverage.

The main goal is to develop a consistent structural–stratigraphic framework for the Plio-Quaternary strata and the Messinian succession of the Levant Margin. Using 2D profiles across regional sections, I identify and characterise the key structural styles, patterns, and deformation features. This framework, adapted from techniques detailed further in Chapter Three, offers a way to examine how structural components relate to one another within the extensional domain of the Levantine Margin.

Why focus on this area? The Levant Margin is one of the least studied extensional regions in the Eastern Mediterranean Sea, yet it also provides one of the world's best natural type locality for examining thin-skin deformation (Garfunkel & Almagor, 1985; Hsü et al., 1973). This dual nature, underexplored yet highly insightful, makes it ideal for testing the limits of 2D seismic interpretation.

The profiles shown here emphasise several key structural features. Previous studies (e.g., Clark & Cartwright, 2009; Cartwright et al., 2012; Nwosu, 2013) have demonstrated that extensional fault interactions are central to gravity-driven tectonics. Building on those findings, this chapter describes the structural styles, vertical fault lengths, and fault polarities observed in the Lebanon dataset.

As you move through this chapter, I encourage you to ask:

- How much of the structural complexity can truly be resolved from 2D profiles alone?
- In what ways might the interpretation shift once I return to the full 3D volume?
- What do these insights tell us about rifted margin tectonics more broadly?

By the end, you should have a clearer understanding of the Levant Margin and a deeper appreciation of the strengths and limitations of 2D seismic analysis in structural geology.

4.2 Developing a Stratigraphic-Structural Framework of 2-D Seismic Profiles

The interpretation process started with selecting a representative 2D seismic profile, exported in a high-resolution graphics format and printed at a scale of 1:25,000 (Fig. 4.1a). The seismic line trends NNE, with water depths from 1525 m to 1906.25 m. The initial interpretation was done manually on the printed section, with colour-coded overlays emphasising key stratigraphic horizons and structural features (Fig. 4.1b–c). Although analogue interpretation is becoming less common in modern seismic workflows, it remains a useful tool for early-stage conceptual mapping, horizon correlation, and educational purposes.

Within this profile, several stratigraphic horizons and numerous fault structures were identified (Fig. 4.1c). The annotated paper sections were then scanned at

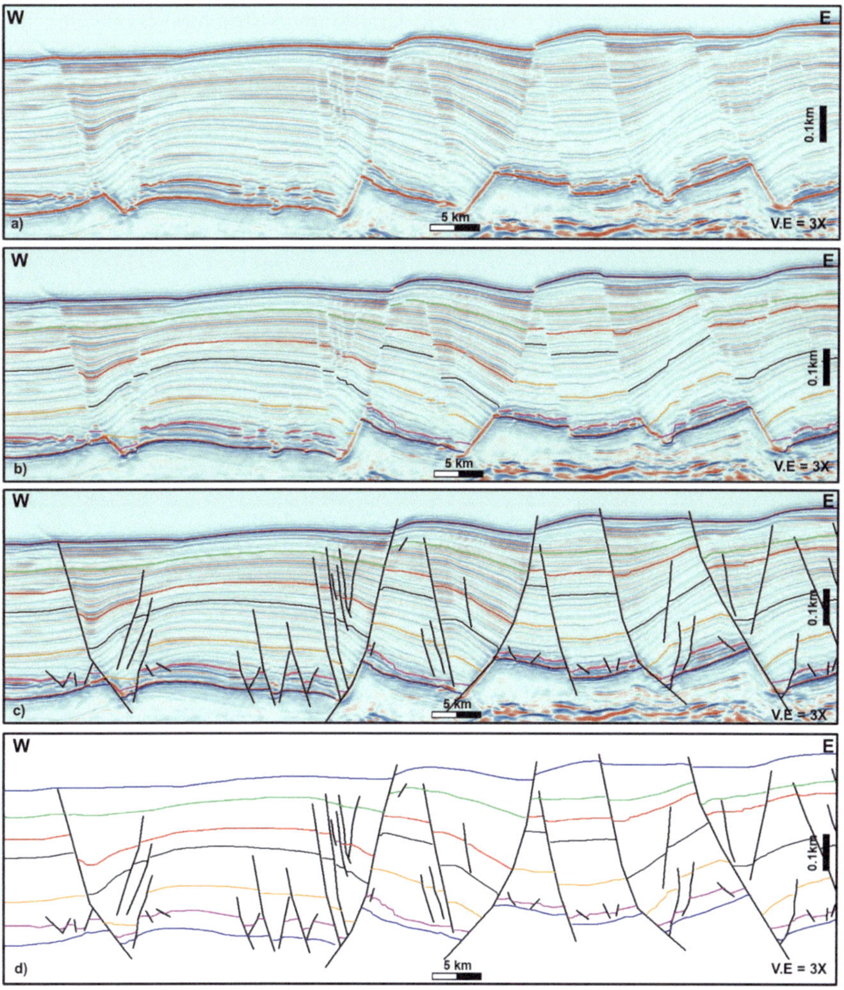

Fig. 4.1 Stratigraphic-structural workflow for 2-D seismic profiles. **a** Uninterpreted seismic section. **b** Horizon interpretation with colour-coded reflectors highlighting key stratigraphic units. **c** Fault interpretation with black lines delineating fault planes cutting through the stratigraphy. **d** Schematic structural model derived from the seismic interpretation, showing fault networks and stratigraphic relationships in a simplified format. Vertical exaggeration (V.E) is set to 3 × to enhance fault visibility

300 dpi and imported into the MOVE software environment, where the integrated mapping tools enabled digital retracing and refinement of the interpreted horizons and faults (Fig. 4.1d). This iterative analogue–digital workflow ensured accuracy while maintaining the clarity of the initial observations.

The same procedure was applied to six regional 2D seismic lines: 21,920, 21,820, 21,700, 21,600, 21,520, and 21,400 (Figs. 4.2, 4.3, 4.4, 4.5, 4.6, 4.7). These profiles collectively provided a consistent stratigraphic–structural framework that underpinned subsequent quantitative analysis and 3-D interpretation.

As you review these profiles, think about the following:

- Why might analogue methods still offer advantages in the earliest stages of interpretation?

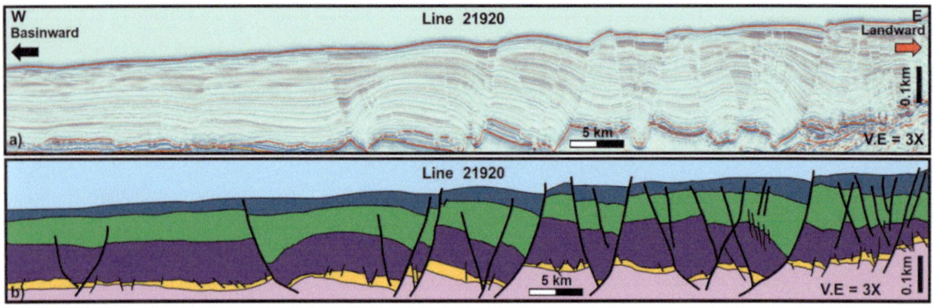

Fig. 4.2 2D seismic profiles showing the **a** uninterpreted section and **b** main stratigraphy and structures of Line 21,920

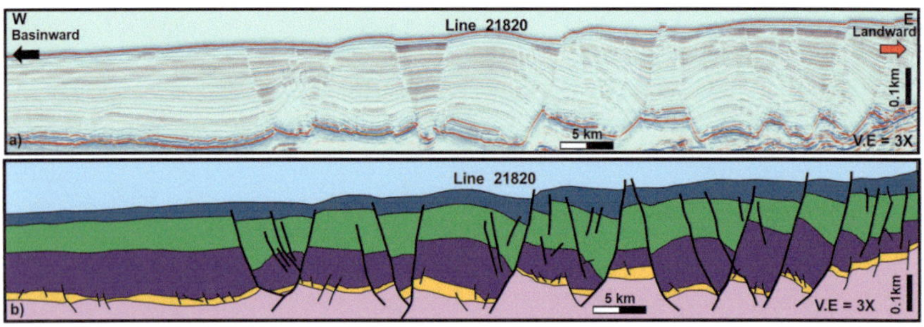

Fig. 4.3 2D Seismic profiles showing: **a** uninterpreted section, and **b** main stratigraphy and structures of Line 21,820

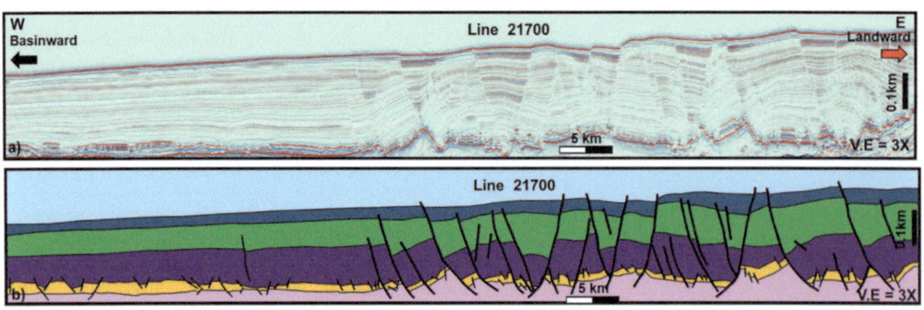

Fig. 4.4 2D Seismic profiles showing the **a** uninterpreted section, and **b** main stratigraphy and structures of Line 21,700

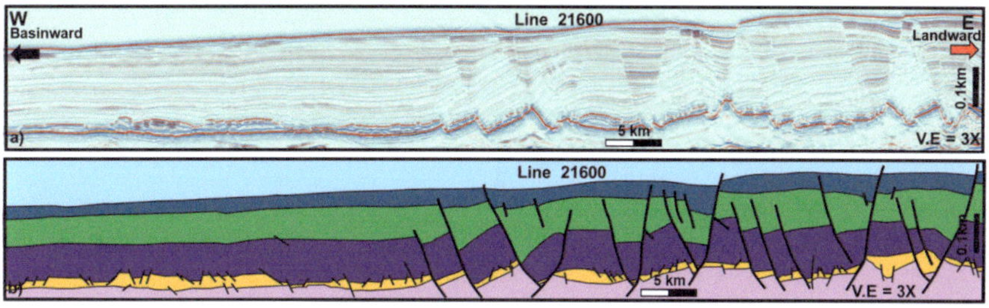

Fig. 4.5 2D Seismic profiles showing the **a** uninterpreted section, and **b** main stratigraphy and structures of Line 21,600

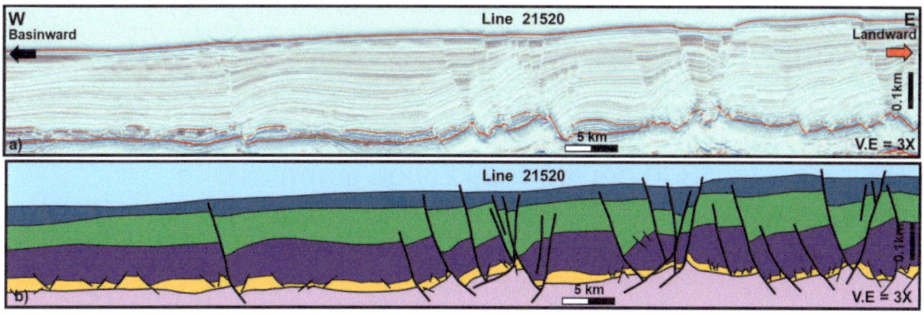

Fig. 4.6 2D Seismic profiles showing the **a** uninterpreted section, and **b** main stratigraphy and structures of Line 21,520

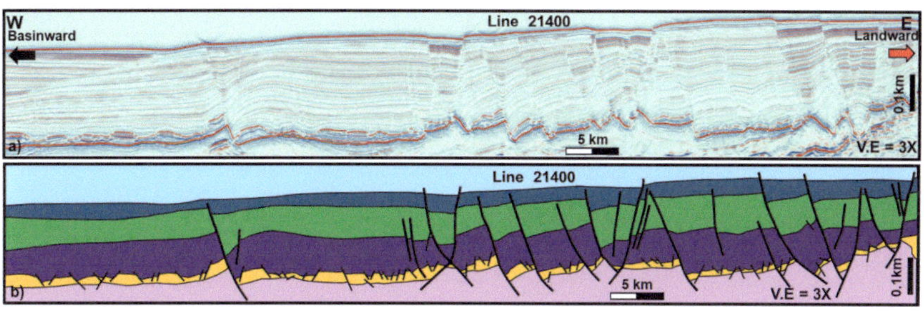

Fig. 4.7 2D Seismic profiles showing the **a** uninterpreted section, and **b** main stratigraphy and structures of Line 21,420

- How does the integration of analogue and digital workflows enhance structural–stratigraphic correlation?
- Which features remain consistent across all six seismic lines, and which ones vary more?

Reflecting on these questions as you examine Figs. 4.1, 4.2, 4.3, 4.4, 4.5, 4.6, 4.7 can help you better understand the balance between traditional and

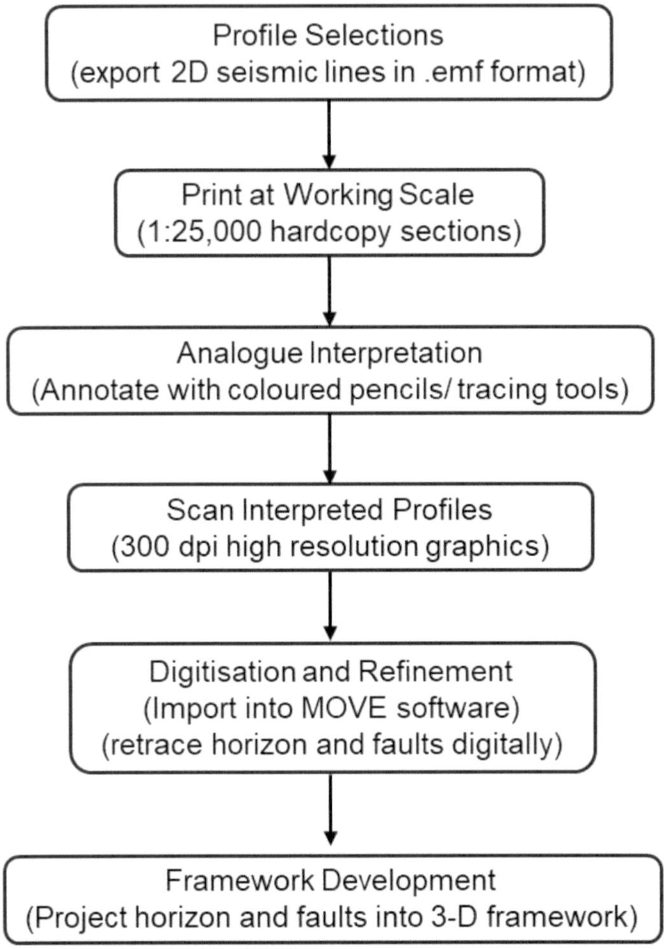

Fig. 4.8 Analogue-to-digital seismic interpretation process

modern approaches to seismic interpretation. A workflow was used to describe the stratigraphic-structural framework of 2-D seismic profiles (Fig. 4.8).

4.3 Seismic–Stratigraphic Units of the Regional 2D Profiles

Interpretation of the regional 2D seismic dataset identified four seismic–stratigraphic units, SP1–SP4. These units were characterised by reflector terminations (truncation, onlap, downlap), amplitude variations, internal reflection configurations, and regional

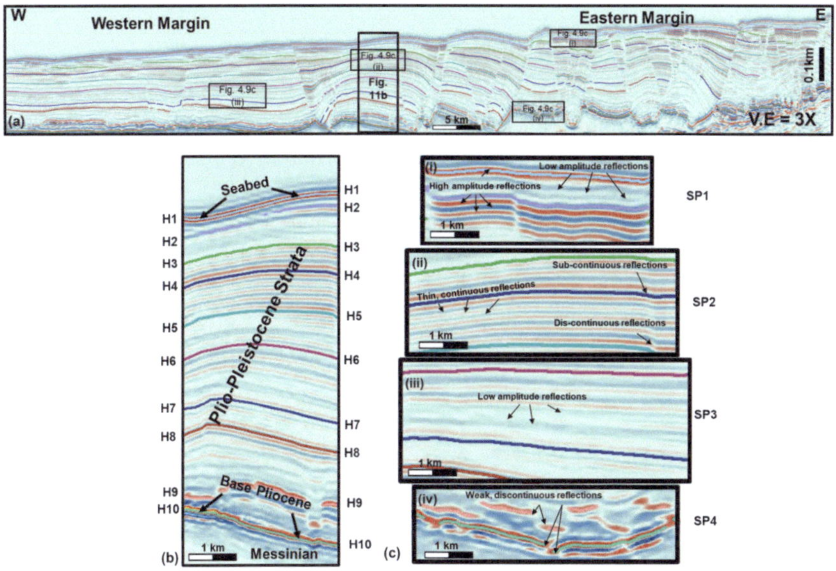

Fig. 4.9 Seismic facies analysis across the western and eastern margins. **a** Seismic profiles showing 10 seismic-stratigraphic horizons, with boxes indicating the sections selected for detailed facies analysis (i) to (iv). **b** Selected sections of Fig. 6.2a, **c** Selected sections of different packages of seismic facies units (i) SP1 showing high- and low-amplitude reflections highlighting lithological variations; (ii) SP2 indicating sub-continuous and thin, continuous reflections, suggesting variations in sedimentary processes; (iii) SP3 showing low-amplitude reflections indicative of homogeneous lithology; (iv) SP4 showing weak, discontinuous reflections corresponding to structural or stratigraphic discontinuities

bounding markers. For clarity, colour coding was used: SP1 (blue), SP2 (green), SP3 (purple), and SP4 (orange).

Interpretation was based on ten seismic horizons (H1–H10), with Line 21,920 serving as the reference profile to illustrate facies architecture representative of the wider dataset (Fig. 4.9a, b).

4.3.1 Unit SP1

SP1 features moderate to high amplitude, parallel reflections with variable continuity, ranging from continuous to discontinuous. This pattern predominantly appears along the eastern margin and parts of the western margin of the profiles. A sigmoidal reflection geometry is observed locally on the western margin (Fig. 4.9c–i).

The average thickness of SP1 is about 395 m. The unit shows no clear signs of syn-kinematic deformation and is therefore considered relatively undisturbed. Two horizons were mapped: H1 (seabed reflector), which appears as a strong, continuous high-amplitude peak, and H2 (Fig. 4.9c–i).

4.3.2 Unit SP2

SP2 consists of low- to medium-amplitude reflections with an average thickness of 593 m. Reflection continuity varies: it remains predominantly continuous in the west but becomes discontinuous towards the east. Compared to SP1, the reflector spacing is closer.

A low-amplitude reflector marks the upper boundary of SP2 (H4), while its base (H6) aligns with a higher-amplitude event (Fig. 4.9c–ii). Internally, the unit exhibits limited syn-kinematic deformation, although local thickness variations occur due to channel–levee complexes that onlap fold crests (Clark & Cartwright, 2009). Three reflectors, H3, H4, and H5, are identified within SP2.

4.3.3 Unit SP3

SP3 is characterised by low- to moderate-amplitude, high-frequency reflections, with an average thickness of 988 m (Fig. 4.9c–iii). The seismic character varies laterally: continuous, parallel reflections are common on the eastern margin, whereas the western margin shows more discontinuous configurations.

Evidence of syn-kinematic deformation is more evident in SP3 than in the overlying units. Horizons H6, H7, and H8 were interpreted to define internal stratification with moderate structural overprinting.

4.3.4 Unit SP4

SP4 forms the lowest stratigraphic package and is characterised by weak-to-medium-amplitude, discontinuous, and chaotic reflections throughout the section (Fig. 4.9c–iv). The unit has an average thickness of 296 m and exhibits poor lateral continuity.

Two reflectors were mapped: H9 and H10, the latter interpreted as the top of the Messinian evaporites, expressed as a strong positive reflection (Table 4.1).

Units SP1–SP3 form the Plio–Pleistocene sequence, with SP4 marking the Base Pliocene. Units SP2–SP4 are difficult to correlate across faults because their reflectors are variable and discontinuous, often traceable only locally. Their thickness and seismic variation reflect depositional changes and structural overprinting.

Table 4.1 Seismic–stratigraphic characteristics of units SP1–SP4 across the regional 2D seismic dataset. Thickness values are average estimates. The character of reflections is described in terms of amplitude, continuity, and internal geometry

Unit	Average thickness (m)	Reflection Character	Boundaries	Deformation evidence
SP1	395	Moderate–high amplitude; parallel; continuous to discontinuous; local sigmoidal geometry (west)	H1 (seabed) and H2	None; relatively undisturbed
SP2	593	Low–medium amplitude; tighter spacing; continuous (west), discontinuous (east)	Upper: H4 (low amplitude); Base: H6 (high amplitude)	Limited; local thickness variations from channel–levee onlap
SP3	988	Low–moderate amplitude; high-frequency; continuous (east), discontinuous (west)	Internal: H6, H7, H8	Moderate; evidence of syn-kinematic deformation
SP4	296	Weak–medium amplitude; discontinuous; chaotic; poor lateral continuity	H9 and H10 (top Messinian evaporites)	Substantial heterogeneity; difficult correlation across faults

4.4 Conclusion

The regional stratigraphic–structural framework of the 2D seismic dataset from the Levantine Margin was developed around two core themes: (1) the stratigraphic–structural interpretation of 2D seismic profiles and (2) the definition of seismic–stratigraphic units. The main conclusions are as follows:

1. Regional framework construction: Six seismic lines (Lines 21,920, 21,820, 21,700, 21,600, 21,520, and 21,400) were selected and interpreted within 2D regional profiles. These lines served as the basis for horizon and fault mapping.
2. Seismic–stratigraphic units: Four seismic facies packages were identified within the interval between the Top Messinian and Plio-Quaternary successions. These units, labelled SP1, SP2, SP3, and SP4, constitute the stratigraphic core of the interpretation.
3. Framework characteristics: They display notable vertical variations (such as thickness, reflector amplitude, stratigraphic stacking) and lateral heterogeneity between the eastern and western margins. These variations reflect both depositional processes and structural modifications throughout the basin.

Taken together, this framework highlights the importance of combining 2D seismic interpretation with stratigraphic correlation to understand the tectono-stratigraphic development of rifted continental margins, like the Levant Basin.

Learning Objectives

After completing this chapter, readers will be able to:

1. Interpret seismic–stratigraphic units using reflector terminations and facies patterns.
2. Correlate stratigraphic units across regional 2-D seismic profiles.
3. Differentiate between continuous, discontinuous, sigmoidal, and chaotic seismic facies.
4. Evaluate the role of analogue methods in early-stage interpretation.
5. Integrate 2-D and 3-D interpretations to build regional stratigraphic frameworks.

References

Bertoni, C., & Cartwright, J. A. (2006). Controls on the basinwide architecture of the late Miocene (Messinian) evaporite on the Levant margin (Eastern Mediterranean). *Sedimentary Geology, 188*, 93–114.

Cartwright, J. A., Trudgill, B. D., & Mansfield, C. S. (1995). Fault growth by segment linkage: An explanation for scatter in maximum displacement and trace length data from the Canyonlands Grabens of SE Utah. *Journal of Structural Geology, 17*(9), 319–1326.

Clark, I. R., & Cartwright, J. A. (2009). Interactions between submarine channel systems and deformation in deepwater fold belts: Examples from the Levant Basin and the Eastern Mediterranean Sea. *Marine and Petroleum Geology, 26*, 1465–1482.

Garfunkel, Z., & Almagor, G. (1985). Geology and structure of the continental margin off northern Israel and the adjacent part of the Levant Basin. *Marine Geology, 2*(2), 105–131.
Hsü, K. J., Ryan, W. B. F., & Cita, M. B. (1973). Late Miocene desiccation of the Mediterranean. *Nature, 242*, 240–244.
Netzeband, G. L., Hübscher, C., & Gajewski, D. (2006). The structural evolution of the Messinian evaporites in the Levantine Basin. *Marine Geology, 230*(3–4), 249–273.
Nwosu, O.: The compartmentalisation of a deepwater fold and thrust belt in the levant. (PhD thesis), Cardiff University, U.K. (2013).

Chapter 5
Understanding Three-Dimensional (3D)/ Map View Workflows of Seismic Data

Abstract This chapter offers a three-dimensional (3D) map-based interpretation of seismic structure, illustrating how 3D seismic datasets expand structural analysis beyond the constraints of two-dimensional (2D) profiles. Using a high-resolution 3D seismic volume from the extensional domain of the Levantine Margin, the chapter details a workflow for integrating 2D interpretations into a cohesive 3D structural–stratigraphic framework. Key stratigraphic horizons, such as the seabed and the Top Messinian surface, are mapped and converted from time to depth, yielding detailed structural maps across a 625 km^2 area. A total of 364 faults are identified, mapped, and classified into four structural domains (FM-1 to FM-4), facilitating quantitative analysis of fault orientation, spatial distribution, and depth-dependent variability. The findings reveal variations in fault frequency with stratigraphic level and highlight complex fault interactions that are difficult to resolve in 2D sections alone. By explicitly connecting 2D seismic lines with the 3D volume, the chapter demonstrates how fault continuity, linkage, and termination can be reliably established. Readers are encouraged to consider how 3D mapping transforms structural interpretation from isolated profiles into an integrated, spatially consistent geological model.

Keywords Depth-structure maps · Fault architecture · Fault segmentation · Structural domains · Fault connectivity

5.1 Introduction

In the previous chapter, I examined 2D seismic profiles and observed that fault structures vary greatly in orientation and frequency across different lines. Did you notice that some faults end suddenly, while others extend laterally with significant continuity? Such variability naturally raises the question: how are these structures connected when I move beyond a single profile?

To address this, I need to consider the scope of three-dimensional (3D) seismic mapping. Unlike 2D lines, 3D datasets allow us to visualise fault architecture in map

K. Chiadikobi, *Advances in Horizon Mapping and Fault Interpretation*,
https://doi.org/10.1007/978-3-032-23917-4_5

view, trace their continuity across the margin, and improve our understanding of how they interact in space. This chapter guides you through that process.

Here, I describe and interpret the structural-architectural framework of a high-resolution 3D seismic dataset acquired from the extensional domain of the Levantine Margin. This region provides an excellent natural test locality: its structures result from large-scale gravity-driven collapse, spreading basinward into the deeper parts of the Levant Basin. Numerous authors (e.g., Ben-Gai et al., 2005; Cartwright & Jackson, 2008; Cartwright et al., 2012; Clark & Cartwright, 2009; Garfunkel, 1998; Gradmann et al., 2005) have highlighted the complexity of this margin, making it an ideal setting for testing structural concepts.

This chapter aims to quantify and characterise the mapped structures within the Plio–Quaternary and Top Messinian intervals. By the end of this chapter, you should be able to recognise how 3D mapping enhances our interpretation compared with 2D sections. As you read, consider: What advantages does 3D mapping offer for interpreting fault architecture that 2D profiles alone cannot provide?

5.2 Structural-Stratigraphical Framework of 3D Seismic Data

Integrating 2D seismic interpretations into a three-dimensional (3D) seismic volume provided the basis for constructing a comprehensive structural–stratigraphical framework. Horizons and faults previously identified from 2D profiles (see Chap. 4) were imported into the simulation tool, enabling full 3D visualisation and spatial correlation.

Within the 3D volume, I mapped key stratigraphic reflectors and fault structures to develop a detailed geological model. The seabed horizon served as the primary reference surface. To achieve this, I created a surface polygon covering approximately 625 km^2, focusing on a structurally complex zone characterised by a high density of intersecting faults (Fig. 5.1a).

The interpreted horizons were then transformed into stratigraphic surfaces. This process allowed for direct 3D inspection and iterative correction of interpretation inconsistencies that could have remained hidden in 2D sections. This method refined our understanding of fault geometries, reflector terminations, and stratigraphic relationships. The contoured surfaces were subsequently converted from two-way time (TWT) to depth using a constant-velocity model of 1975 m/s, resulting in a series of depth-structure maps (Figs. 5.2, 5.3, 5.4 and 5.5).

Across the seismic volume, I mapped and interpreted 364 faults. These faults crosscut stratigraphic horizons and are visualised in 2D map views. To facilitate structural analysis, I grouped the faults into four domains (FM-1 to FM-4), each exhibiting distinct geometric and kinematic styles (Figs. 5.6, 5.7, 5.8 and 5.9). The fault maps were essential for quantifying vertical segmentation and lateral variability within fault systems.

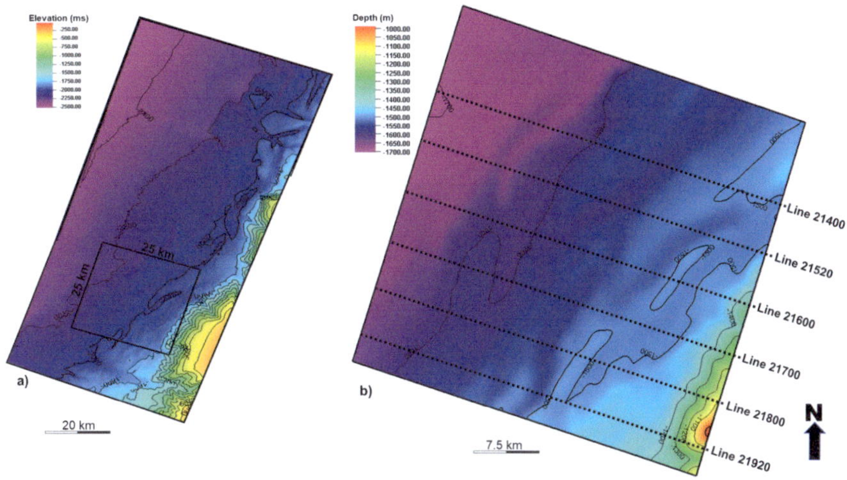

Fig. 5.1 **a** Regional time-structure map of the seismic survey, with the area of interest outlined. **b** Depth-structure map of the selected area (625 km^2), showing key seismic lines (21,520, 21,600, 21,700, 21,800, and 21,920) used for seismic interpretation

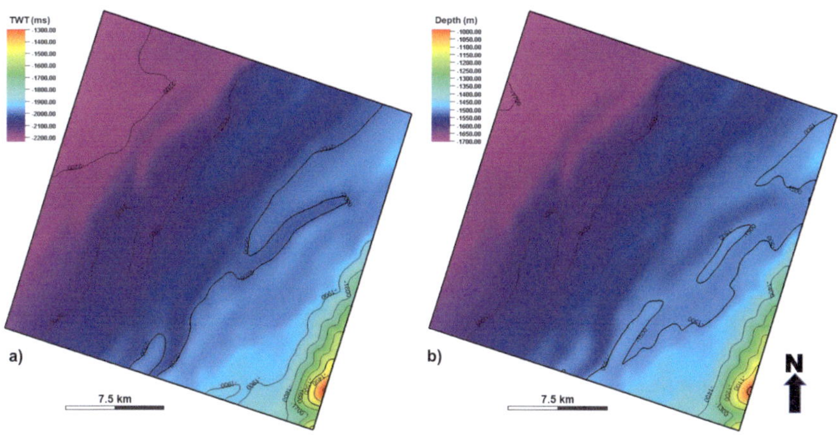

Fig. 5.2 Horizon maps showing **a** Time-Structure (TWT) and **b** Depth-Structure (m) of the Seabed

To ensure interpretive confidence, I employed a rigorous validation workflow. This included attribute-based analysis, geometric assessment, correlation, and geological plausibility testing. These checks were especially important for resolving subtle, small-scale fault structures near the seismic resolution limit and for differentiating genuine geological features from artefacts introduced during acquisition or processing.

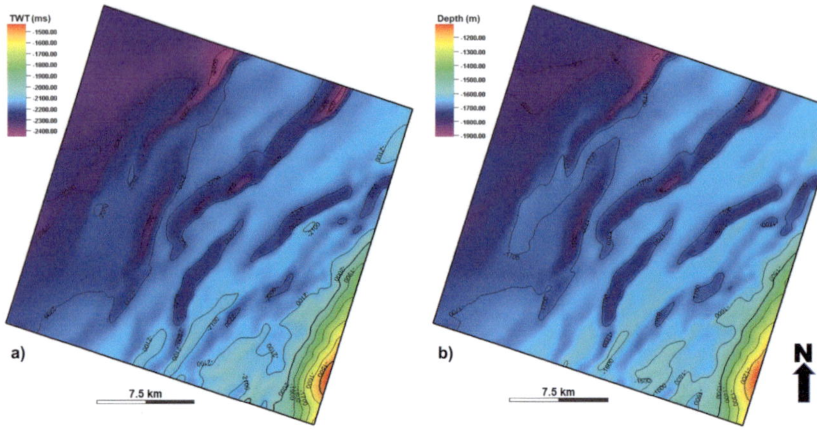

Fig. 5.3 Horizon maps showing **a** Time-Structure (TWT) and **b** Depth-Structure (m) of Horizon 3

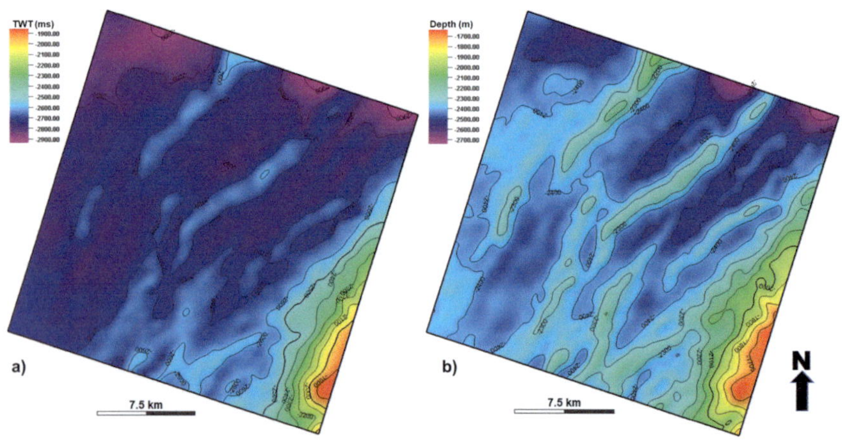

Fig. 5.4 Horizon maps showing **a** Time-Structure (TWT) and **b** Depth-Structure (m) of Horizon 7

Moving from 2D to 3D seismic analysis enhances understanding of fault connectivity, continuity, and stratigraphy. In 2D, faults may look isolated, making linkage difficult. Stratigraphic reflectors that seem broken in one section can be traced throughout the volume, revealing their true shapes. Essentially, 3D analysis turns isolated snapshots into a unified, spatial interpretation.

Most faults mapped within the 3D seismic volume strike NNE–SSW, consistent with the survey orientation. Notably, the number of mapped faults increases with

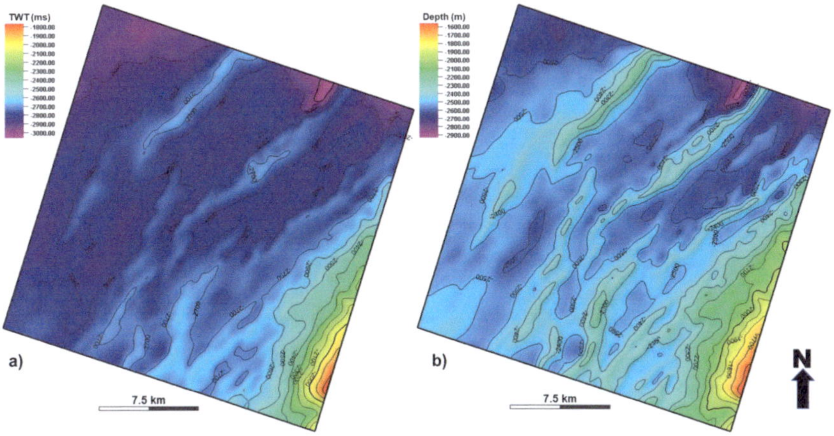

Fig. 5.5 Horizon maps showing **a** Time-Structure (TWT) and **b** Depth-Structure (m) of the Top Messinian

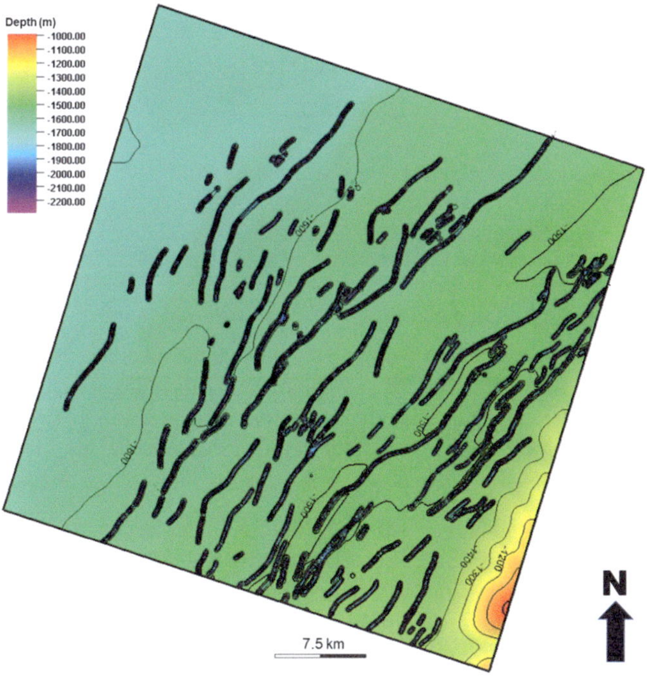

Fig. 5.6 2D map of fault arrays on stratigraphic horizons of FM-1

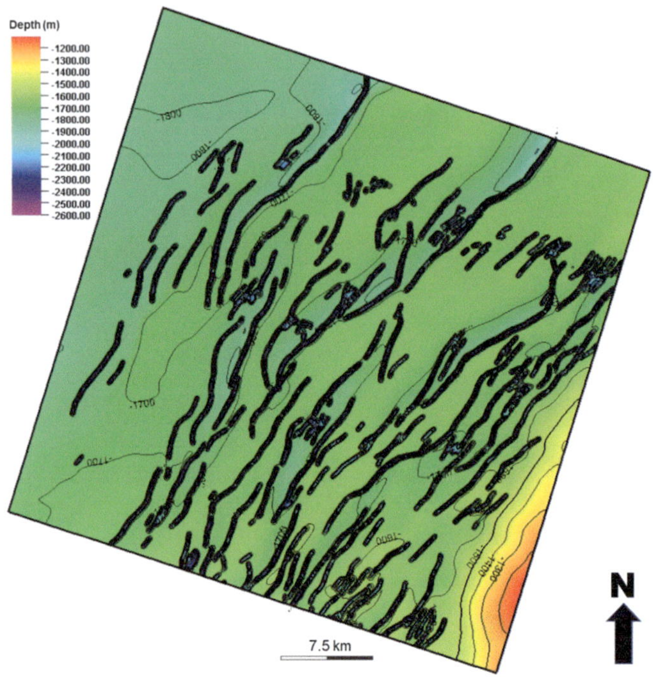

Fig. 5.7 2D map of fault arrays on stratigraphic horizons of FM-2

depth, indicating progressive fault segmentation or activation in deeper stratigraphic intervals.

The 2D structural maps offer an additional perspective. They reveal a gradual decrease in fault numbers basinward, with most faults striking obliquely (SSE–SSW) relative to the dominant NNE–SSW fault set. This obliquity emphasises local variations in the stress regime and potential fault reactivation.

Fault population variability is further demonstrated by the four mapped fault domains (FM-1 to FM-4; Figs. 5.6, 5.7, 5.8 and 5.9):

- FM-1 (Seabed): Depth ranges from ~1300 to 2200 m. Fault frequency is highest between 2000 and 2100 m and lowest at shallower depths of <1900 m (Fig. 5.6).
- FM-2: Elevations range from ~1500 to 2500 m below sea level. Fault frequency peaks at 2100–2200 m, with the lowest frequencies recorded below 2300 m (Fig. 5.7).
- FM-3: Elevations range from ~1900 to 2900 m. The fault population peaks at 2700–2800 m, whereas the lowest counts occur outside this range (Fig. 5.8).
- FM-4 (Top Messinian): Elevations range from ~ 1800 to 3100 m. Fault frequency is highest between 2700 and 2800 m and lowest between 2800 and 2900 m (Fig. 5.9).

These observations confirm that fault distribution is uneven, with fault frequency varying by depth, stratigraphic interval, and basin location.

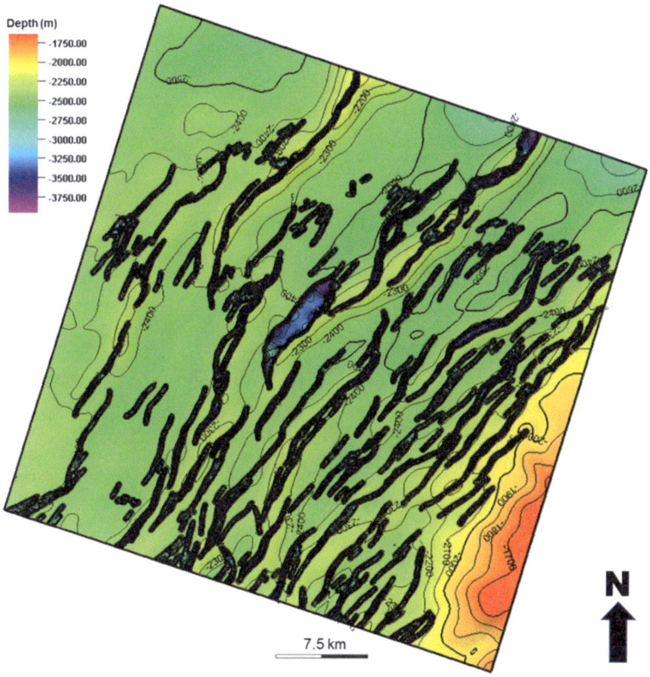

Fig. 5.8 2D map of fault arrays on stratigraphic horizons of FM-3

5.3 Integration of 2D and 3D Seismic Interpretations

The 3D seismic interpretation was linked to the selected 2D seismic sections (e.g., Lines 21,920, 21,800, 21,700, 21,600, 21,520, and 21,400; see Chap. 4). Within the 3D volume, these 2D profiles appear as intersecting lines, allowing direct visualisation of how fault components identified in profile view connect in map view. This integration emphasises fault connectivity across profiles and reinforces the benefits of interpreting structural architecture in three dimensions.

In the next chapter, I analyse and quantify variations in fault geometries across different horizon maps, based on structural styles and dip directions.

Linking 2D profiles to the 3D seismic cube improves interpretive confidence by allowing faults in cross-sections to be traced on maps. 2D shows vertical detail, but 3D confirms fault continuity, linkages, and terminations, reducing misinterpretation, like confusing listric with planar faults, by providing a full spatial setting.

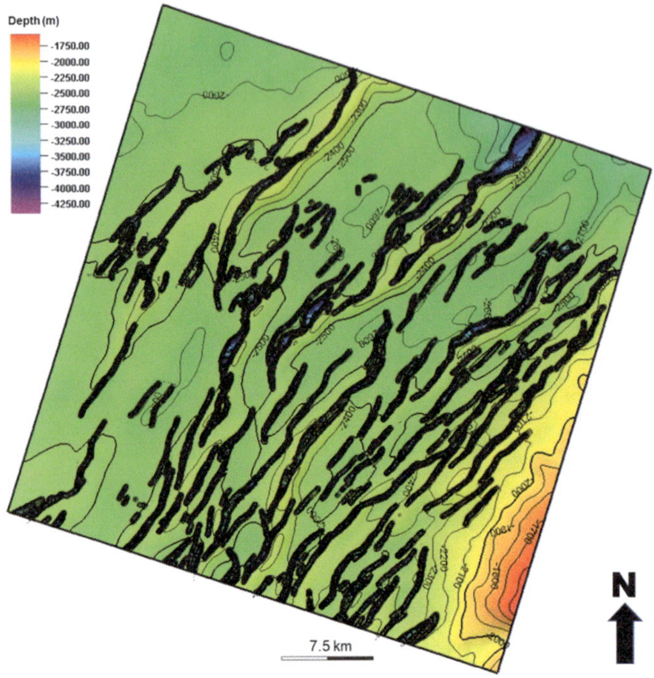

Fig. 5.9 2D map of fault arrays on stratigraphic horizons of FM-4

5.4 Conclusion

This chapter shows that three-dimensional (3D) seismic mapping is essential for understanding the true shape, continuity, and interaction of fault systems in structurally complex extensional margins. By combining interpretations from two-dimensional (2D) seismic profiles into a high-resolution 3D seismic volume, isolated or unclear structures seen in profile view are turned into a consistent spatial framework. This method allows faults to be traced sideways, connected across profiles, and assessed within the full map-view layout.

The mapping of key stratigraphic horizons and the interpretation of 364 faults across a 625 km^2 area show that fault distribution, orientation, and frequency vary considerably with depth and stratigraphic level. Grouping faults into four structural domains (FM-1 to FM-4) offers a solid basis for quantifying fault population variability and recognising depth-dependent segmentation patterns. These findings confirm that faulting within the Levantine Margin is neither uniform nor random, but organised and strongly influenced by stratigraphic position and gravitational tectonic processes.

Importantly, combining 2D and 3D seismic interpretations improves interpretational confidence by reducing geometric uncertainty and minimising misclassification of fault styles. While 2D profiles retain essential vertical detail, 3D mapping reveals fault connectivity and interactions that cannot be reliably inferred from profiles alone.

Learning Objectives

After completing this chapter, readers will be able to:

1. Identify faults using reflector offsets, dip changes, amplitude variations, and monocline geometries.
2. Apply systematic fault-mapping workflows across 3D seismic volumes.
3. Distinguish between genuine fault geometries and seismic artefacts.
4. Assess fault detectability limits imposed by seismic resolution.
5. Construct internally consistent fault networks through rigorous quality control.

References

Ben-Gai, Y., Ben-Avraham, Z., & Garfunkel, Z. (2005). Seismic reflection study of the continental margin off northern Israel. *Tectonophysics, 398*(1–2), 103–123.

Cartwright, J., & Jackson, M. P. A. (2008). Initiation of gravitational collapse of an evaporite basin margin: The Messinian saline giant, Levant Basin. *Geological Society of America Bulletin, 120*(3–4), 399–413.

Cartwright, J., Jackson, M. P. A., & Dooley, T. (2012). Stratigraphic controls on gravity-driven deformation in passive margins. *Geological Society, London, Special Publications, 372*, 1–15.

Clark, J. A., & Cartwright, J. A. (2009). Interaction between deeply seated and shallow normal faults in the Levant Basin, Eastern Mediterranean. *Basin Research, 21*(6), 703–721.

Garfunkel, Z. (1998). Constraints on the origin and history of the Eastern Mediterranean basin. *Tectonophysics, 298*(1–3), 5–35.

Gradmann, S., Netzeband, G. L., Hübscher, C., & Gajewski, D. (2005). Seismic characterisation of salt tectonics in the Levant Basin, Eastern Mediterranean. *Marine and Petroleum Geology, 22*(6–7), 701–715.

Chapter 6
Description and Quantification of Fault Geometries

Abstract This chapter outlines the fault geometries, structural styles, and polarity patterns that shape the extensional architecture of the Levantine continental margin, based on an integrated interpretation of regional 2D seismic profiles and high-resolution 3D fault maps. It demonstrates how closely spaced normal faults, usually organised into NE–SW-trending arrays, influence vertical and lateral variations in basin structure. The chapter identifies four main structural styles (apparent planar faults, apparent listric faults, block faulting, and conjugate fault systems). It explains how each one contributes to margin evolution through gravity-driven tectonics. Quantitative analysis emphasises the predominance of planar faults in Plio-Quaternary strata, the distinctive curved geometries and rollover structures of listric faults, and the formation of graben–horst and rotated block systems. Conjugate faults are categorised into V-, Y-, and X-shaped geometries, highlighting complex fault interactions and linkage zones. Fault polarity analysis further differentiates synthetic and antithetic faults, showing clear lateral variations in strain partitioning across the margin. The chapter concludes by illustrating how fault attribute maps, heave distribution, dip azimuth, and dip angle serve as effective visual tools for linking fault geometry to regional stress regimes. Readers are encouraged to relate fault form, orientation, and polarity to basin-scale tectonic processes and margin evolution.

Keywords Structural styles · Block faulting · Conjugate faults · Fault polarity · Fault arrays · Fault attributes

6.1 Introduction

In this chapter, I explore the variety of structural styles shown in both 2D seismic profiles and 3D fault maps. As you read, continue asking yourself: How do different fault geometries influence basin evolution, and why are their orientations significant?

When analysing the regional 2D profiles, observe how closely spaced normal faults dominate the architecture, with fault spacing typically ranging from 0.25 to 2 km in the larger structures. These faults show varying displacement directions,

K. Chiadikobi, *Advances in Horizon Mapping and Fault Interpretation*,
https://doi.org/10.1007/978-3-032-23917-4_6

dipping basinward (SW) or landward (SE), and often form fault arrays that control the overall structural configuration. Regional 2D profiles illustrate the vertical variations in fault geometry.

Unlike the 2D sections, the 3D view highlights lateral variations in fault geometry. Notice how the faults, despite their different styles, share a common feature: they are closely spaced normal faults arranged in systematic arrays. Most of these arrays strike or trend mainly NE–SW, reflecting the regional stress regime that shaped the basin.

6.2 Fault Lengths

Fault lengths are most accurately identified in map view, which outlines the lateral extent of a fault trace, measured directly on horizon maps, and serves as a primary indicator of fault growth, linkage, and strain distribution. Fault traces interpreted from 2D horizon maps display a wide range of lateral lengths, reflecting differences in fault initiation density and propagation behaviour across the mapped stratigraphic surfaces. For consistency and comparison, fault lengths are categorised into four groups: very small faults (0–99 m), small faults (100–249 m), medium faults (250–499 m), and long faults (500–1000 m). These categories are documented across all mapped horizons (see Fig. 6.1).

6.2.1 Very Small Faults

Extremely small faults are present on all four horizons and form the most common fault population in the Levantine Margin. In map view, these faults are typically straight or planar and exhibit limited lateral continuity, suggesting restricted propagation lengths. Their orientations in map view are multi-directional, predominantly aligned NNW–SSE and NNE–SSW, indicating fault nucleation under a stress field that permits multiple extension directions rather than a single, dominant trend. The simple geometry and short trace lengths of these faults are consistent with early-stage fault development or the segmentation of strain into many small structures.

The distribution of very small faults varies among fault models, showing systematic spatial differences in fault density. FM-2 has the highest concentration, with 32 faults identified, whilst FM-3 and FM-4 each have 11 (see Table 6.1 and Fig. 6.2). This contrast suggests that fault initiation was more intense or better preserved in FM-2, whereas FM-3 and FM-4 record fewer short fault traces, possibly due to fault linkage, abandonment, or reduced strain accommodation at these horizons. Such variations emphasise that fault length populations are not uniform throughout the stratigraphic succession but respond to changes in mechanical layering, stress distribution, and fault interaction over time.

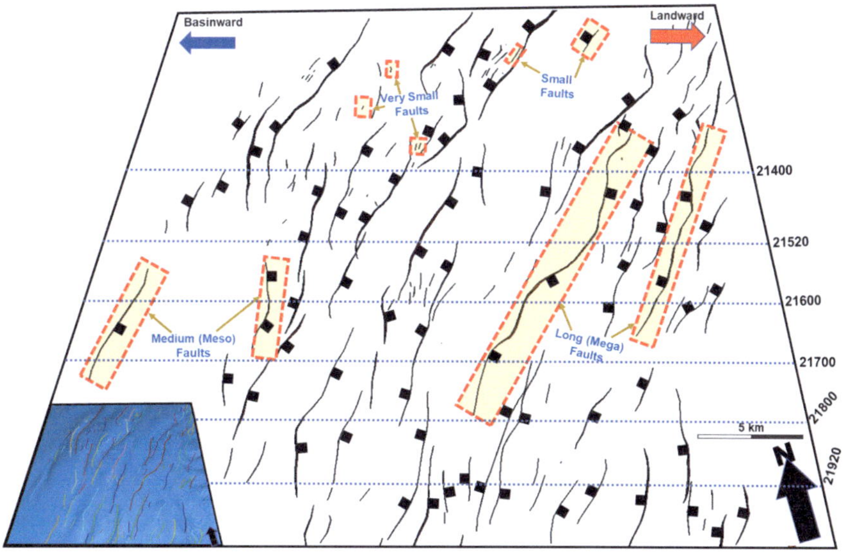

Fig. 6.1 Map-view diagram illustrating fault-length classification and spatial organisation across the Levantine Margin. Very small and small faults occur as short, closely spaced segments distributed throughout the fault network. Medium (meso-scale) faults show greater continuity and localised clustering. Long (mega-scale) faults form laterally extensive, rift-defining structures that dominate the structural framework. Blue dashed lines mark regional seismic profiles (e.g. Lines 21,400–21,920), arrows indicate basinward and landward directions, and shaded envelopes highlight representative fault-length classes

Table 6.1 Summarises the frequency distribution of fault-length classes identified across the four fault maps (FM-1 to FM-4), providing a quantitative basis for assessing fault population evolution and strain partitioning

	V. small faults	Small faults	Medium faults	Mega faults
FM-1	14	60	48	28
FM-2	32	100	56	23
FM-3	11	117	55	23
FM-4	11	102	60	26

6.2.2 *Small Faults*

Small faults are evenly distributed across the mapped horizons, forming a widespread fault population. In map view, these faults appear both straight and gently curved, indicating differing propagation paths and local interactions with nearby structures (see Fig. 6.1). Their extensive spatial distribution suggests that strain was widely distributed rather than concentrated along a few structures. Most small faults strike in NNE–NNW directions, aligned with the main fault sets (see Fig. 6.1). This consistent

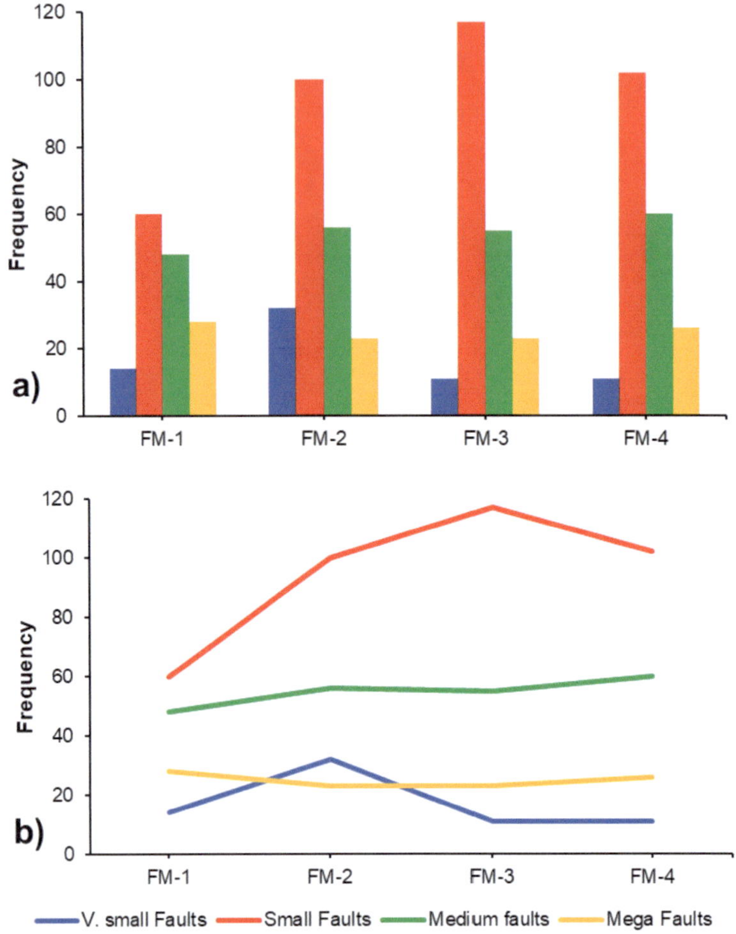

Fig. 6.2 Frequency distribution of fault-length classes across the four fault maps (FM-1–FM-4). **a** Histogram illustrating the relative abundance of very small, small, medium, and long faults, emphasising the dominance of small faults within the fault population. **b** Line diagrams depicting variations in fault-length frequencies between fault maps, highlighting systematic changes in fault populations and the progressive reorganisation of extensional strain through the stratigraphic succession

orientation indicates that these small faults formed under the same regional stress regime that influenced the development of larger faults, and that they are a vital part of the extensional fault network rather than isolated features.

Quantitatively, small faults are most common in FM-3, with 117 faults identified. In contrast, FM-1 records the fewest faults, with 60 (see Table 6.1 and Fig. 6.2). This variation reflects differences in fault nucleation density and fault preservation between models and highlights the sensitivity of small-fault populations to stratigraphic position and local mechanical conditions.

6.2.3 Medium Faults

Medium (meso-scale) faults display sinuous to gently curvilinear traces on maps and mainly strike NNE–SSW, aligning with the primary extensional fault orientation (see Fig. 6.1). Their curved shapes indicate more advanced lateral propagation than smaller faults and suggest increasing interaction with nearby structures and local stress disturbances during fault development.

These medium-length faults are consistently and uniformly distributed across all mapped horizons, indicating that meso-scale fault development was an ongoing process throughout the stratigraphic succession rather than being confined to a single structural level. Their widespread presence emphasises their role as an intermediate fault population linking short, early-stage faults to longer, more continuous structures.

In terms of abundance, FM-4 has the highest number of medium faults, with 60 identified structures, while FM-1 has the lowest, with 48 faults (see Table 6.1 and Fig. 6.2). This systematic variation suggests progressive fault growth and reactivation toward younger horizons, and improved preservation of mesoscale faults in FM-4 compared to deeper levels.

6.2.4 Long Faults

Long (mega-scale) faults appear on the map as sinuous, laterally continuous structures that strike NNE–SSW, shaping the main structural trend of the rift system. Their curved paths indicate extended lateral growth and ongoing displacement accumulation, setting them apart from shorter fault groups (see Fig. 6.1). These faults typically act as rift-bounding structures, with fault polarity shifting from west to east along the rift axis. This alternating polarity produces a symmetrically to mildly asymmetrically organised rift architecture and indicates a systematic organisation of extension across the basin. In map view, long faults exhibit a regular spacing pattern that is consistently observed across the mapped horizons.

The spacing between adjacent mega-faults is not random; it is influenced by the mechanical thickness of the seismically defined facies units. Thicker, mechanically coherent units tend to promote wider fault spacing, while thinner or mechanically weaker units facilitate more closely spaced structures. This relationship highlights the primary role of mechanical stratigraphy in controlling fault localisation at the rift scale.

In terms of abundance, FM-1 has the highest number of long faults, with 28 identified mega-faults, while FM-2 and FM-3 record the fewest, each with 23 faults (see Table 6.1 and Fig. 6.2). These variations suggest differences in fault preservation, linkage, or reactivation between fault models, with deeper or earlier horizons capturing a more complete record of through-going rift-bounding structures. Analysis of the fault maps indicates that fault lengths vary systematically along strike, with prominent lateral variations from SSW to NNW. This lateral variability highlights shifts in fault propagation, interaction, and linkage along the rifted margin rather than consistent fault development throughout the basin.

Quantitative analysis shows a distinct temporal pattern in fault abundance. The total number of faults rises sharply from FM-1 to FM-2, indicating increased fault nucleation and strain localisation. From FM-2 to FM-4, fault numbers gradually decline (see Table 6.1), suggesting either fault linkage into fewer, longer structures, selective reactivation, or reduced preservation of smaller faults at higher stratigraphic levels.

Fault length is a vital parameter because it records the history of deformation and fault growth along the Levantine Margin. Variations in fault length offer insights into the timing, magnitude, and style of extensional processes that have shaped the margin.

Fault distribution within the Plio–Quaternary (post-Messinian) units is strongly affected by sediment loading and post-depositional salt deformation. Gradual sediment accumulation increases differential loading, while salt mobility alters local stress fields, collectively determining where faults originate, how far they extend, and which structures remain active over time.

Small faults dominate the fault population, evenly distributed across horizons, with consistent multidirectional trends and varying in abundance between models, revealing early fault development and extensional strain organisation. Medium faults have sinuous geometries, NNE–SSW strikes, and uniform horizon distribution, marking a transitional population that documents their growth and structural maturation. Long (mega-scale) faults shape rift frameworks, show alternating polarity and spacing, influenced by seismic facies thickness, offering insights into large-scale rift architecture and strain partitioning.

6.3 Structural Styles

The 2D regional seismic profiles and 3D fault maps reveal distinct structural styles: (i) Apparent planar faults (faults with relatively straight traces cutting through stratigraphy). (ii) Apparent listric faults (faults curving downward into the subsurface). (iii) Block faulting (expressed as grabens, horsts, and tilted fault blocks). (iv) Conjugate faults (forming intersecting X-, Y-, or V-shaped geometries).

6.3.1 Apparent Planar Faults

Apparent planar faults are characterised by straight, steeply dipping fault traces of various lengths (see Fig. 6.3a). These features are widely recognised in extensional terranes and are linked to the early stages of continental rifting (e.g. Gibbs, 1984; Nwosu, 2013; Shelton, 1984). On the continental slope, apparent planar faults often form linear arrays trending NE–SW. Within the stratigraphic framework, they are most prominent in the Plio-Quaternary units (Post-Messinian).

The 2D seismic profiles illustrate the distribution of these faults. They are visible along the eastern and western edges of Lines 21,920, 21,820, 21,700, 21,600, 21,520, and 21,400 (see Chap. 4). A quantitative analysis shows that Line 21,700 has the highest number of apparent planar faults (97), while Line 21,820 has the fewest (51) (see Fig. 6.4).

In the map view, apparent planar faults mainly define the structural style, with a total of 333 faults. Their geometries range from straight to sinuous, and their orientations mostly strike NNE–SSW (see Fig. 6.4).

Apparent planar faults are straight, steeply dipping, and most common in Plio-Quaternary units. They align in NE–SW linear arrays, consistent with the regional stress regime.

6.3.2 Apparent Listric Faults

Apparent listric faults are characterised by their curved geometry and concave-upward form. These are normal faults in which the dip angle decreases with depth. The upper fault segment is typically steeply dipping but gradually flattens downward or merges into a low-angle detachment surface (see Fig. 6.3b). Within the stratigraphic framework, apparent listric faults mainly occur in the Plio-Quaternary units (Post-Messinian). Their lengths vary, and they often show systematic relationships with regional detachment surfaces.

On the 2D seismic profiles, apparent listric faults are most clearly seen along the eastern margins of lines 21,920, 21,820, 21,700, 21,600, 21,520, and 21,400 (see Chap. 4). Quantitative analysis shows distinct differences in fault abundance, with the highest values on lines 21,700 and 21,520, and the lowest on line 21,600 (see Fig. 6.4).

In the map view, apparent listric faults appear as long, sinuous structures with lenticular fault tips (see Fig. 6.5). Quantitative analysis reveals a fault population of 31 faults. Their morphology and distribution suggest that listric faults accommodate extension not only through slip along the fault plane but also via interactions between deformational processes in both the hanging wall and footwall. These interactions

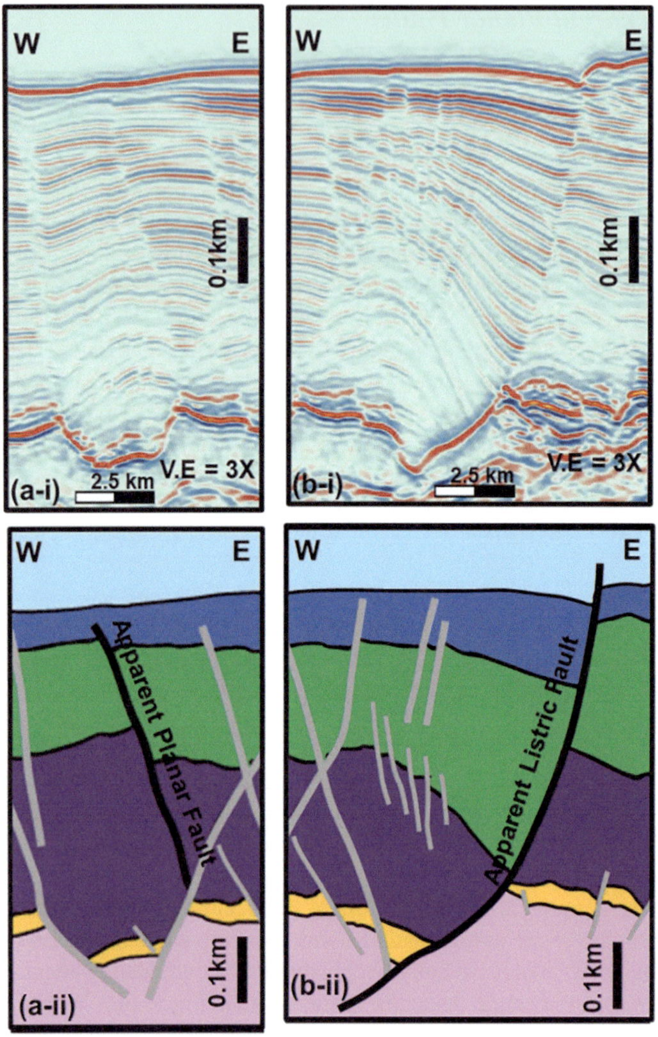

Fig. 6.3 Seismic sections showing Apparent Planar and Apparent Listric Faults. Sections (**a**-i) and (**b**-i) depict uninterpreted seismic profiles. The geoseismic section shows the interpreted faults: (**a**-ii) Apparent Planar fault, (**b**-ii) Apparent Listric fault

provide valuable insights into how gravity-driven tectonics influence the development and growth of listric fault systems on the Levantine Margin. The conceptualised model of the apparent planar and listric faults illustrates the pattern and orientation of their fault systems (see Fig. 6.6).

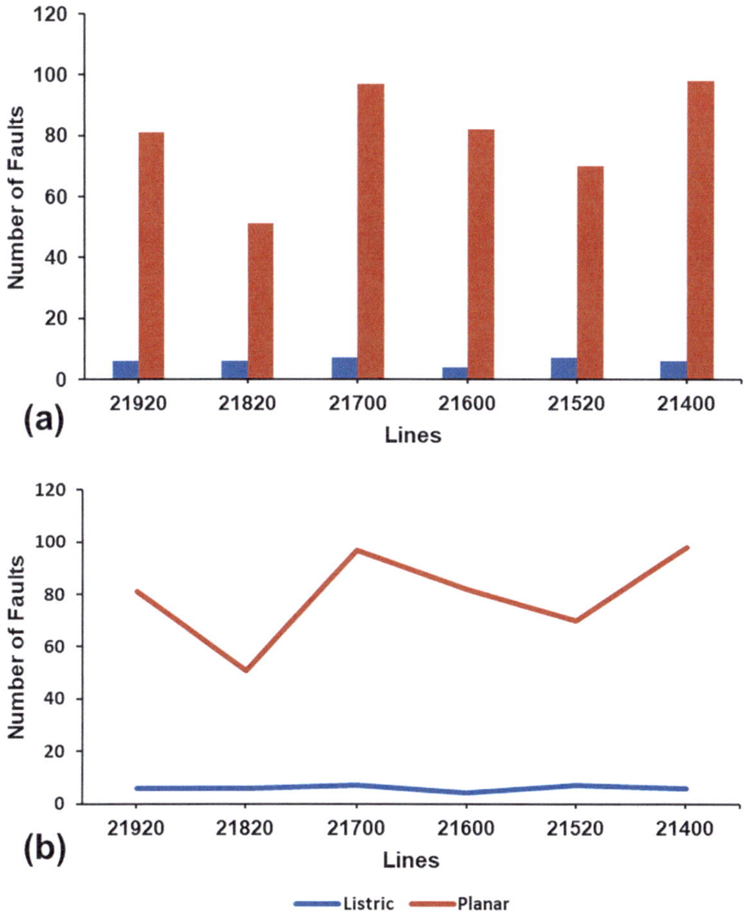

Fig. 6.4 Distribution of listric and planar faults on seismic profiles along Lines 21,920, 21,820, 21,700, 21,600, 21,520, and 21,400. **a** Bar chart showing the relative proportions of listric and planar faults. **b** Line chart illustrating the variation in the ratio of listric to planar faults

On 2D seismic profiles, apparent listric faults exhibit a concave-upward shape, with steep upper parts flattening into detachments. They appear sinuous with lenticular tips, with a total of 31 faults, linking gravity-driven deformation and hanging-wall–footwall interactions and highlighting their role in margin-scale extensional tectonics.

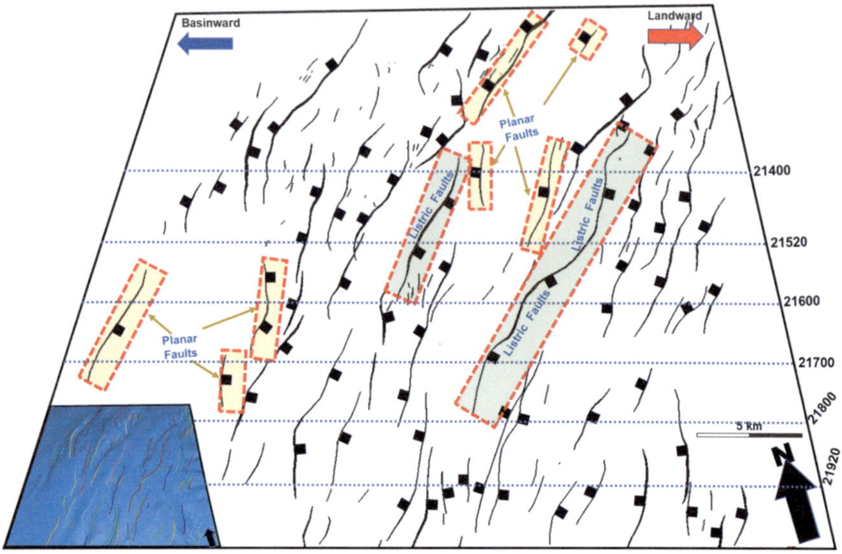

Fig. 6.5 Faul polygon illustrating the spatial distribution and geometric differences between apparent planar and apparent listric normal faults across the Levantine Margin. Planar faults appear as relatively short to medium-length straight segments aligned in NE–SW-trending arrays, while listric faults form longer, sinuous structures that flatten basinward and dominate the architecture at the margin scale. Red dashed envelopes highlight representative fault examples, blue dashed lines indicate regional seismic profiles (Lines 21,400–21,700), and arrows show basinward and landward directions. The diagram emphasises how fault geometry varies laterally and how planar and listric faults coexist within a gravity-driven extensional system

6.3.3 Block Faulting (Grabens, Horsts, and Tilted Faults)

Block faulting on the Levant Margin is a characteristic feature of extensional tectonics, resulting from the interaction of high- to low-angle planar normal faults. These structures develop as the continental margin reacts to localised rift-related stresses. Two main categories of fault blocks are identified:

i. Uplifted blocks bounded by two planar faults, expressed as grabens and horsts.
ii. Tilted blocks, formed by rotation along a single dominant planar fault.

i. *Grabens and Horsts*

Grabens and horsts form where two opposing planar normal faults dip either towards each other (e.g. converging; see Fig. 6.7a) or away from each other (e.g. diverging; see Fig. 6.7b).

On 2D seismic profiles, these structures appear asymmetric, with fault flanks exhibiting two distinct modes:

- Steepening dips with increasing depth, linked to large-scale, widely spaced faults that traverse the Plio-Quaternary and Top Messinian units.

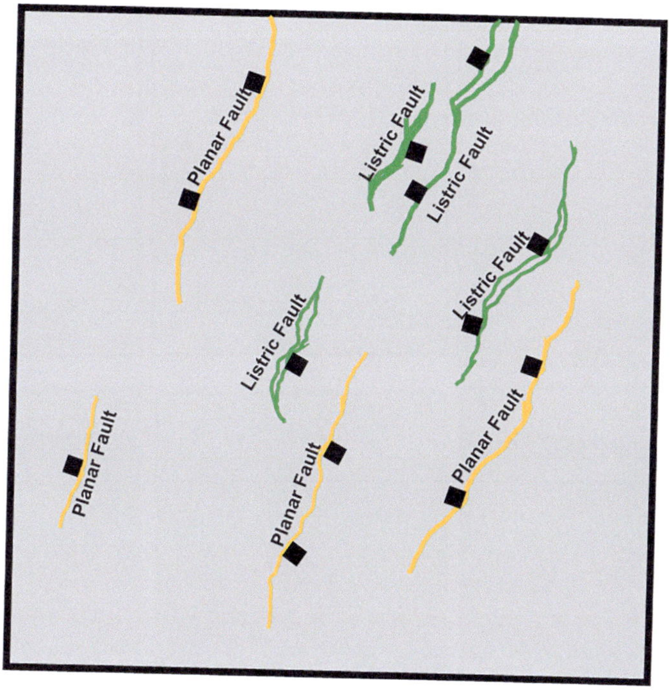

Fig. 6.6 Conceptualised map view model of apparent planar and listric normal faults

- Shallow dips, typical of small-scale, closely spaced faults, are primarily confined to the Top Messinian unit.

Many graben–horst pairs are observed within the Top Messinian, reflecting repeated local extensional events.

In map view, graben and horst structures appear as surface expressions of convergent and divergent conjugate fault systems (see Fig. 6.8). The conceptualised model of graben and horst is shown in Fig. 6.12a.

ii. Tilted Block Faulting

Tilted-block faulting reflects a process of uniform rotational extension, in which fault blocks rotate along steeply dipping planar normal faults. Tilted-block geometries are particularly evident in seismic facies SP2 and SP3 (Fig. 6.7c). Their distribution shows apparent spatial variation:

- Eastern margins: present on Lines 21,920, 21,700, and 21,600 (see Table 6.2).
- Western margins: prominent on Lines 21,820 and 21,700 (see Table 6.2).

This variation suggests strong lateral controls on fault propagation; most likely linked to differential loading and margin-scale detachment geometry.

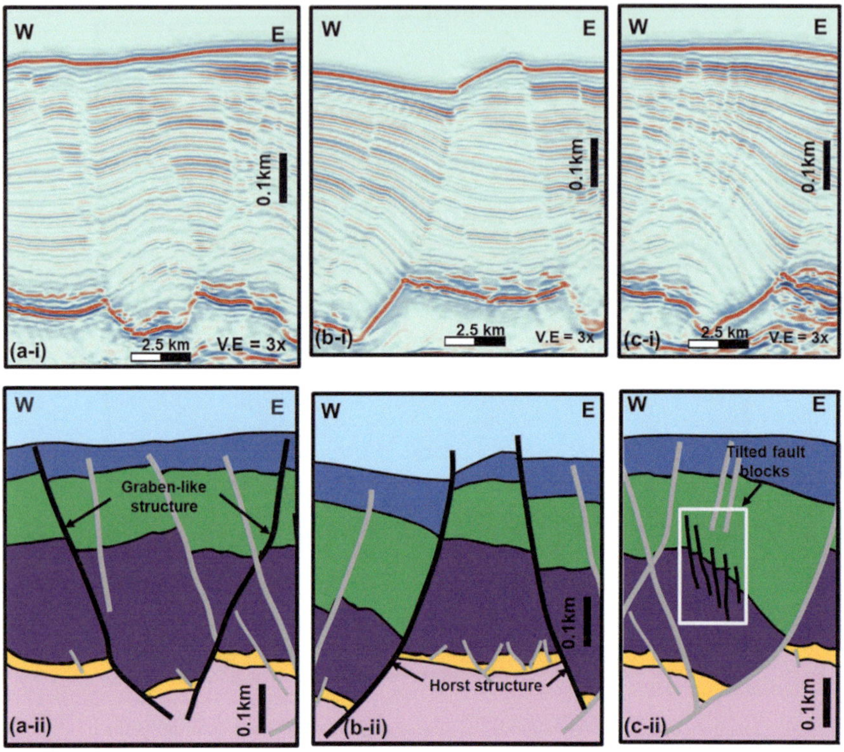

Fig. 6.7 Seismic sections illustrating block faulting. Sections (**a**-i), (**b**-i), and (**c**-i) show uninterpreted seismic profiles. The geoseismic section displays the interpreted faults: **a**-ii Graben, **b**-ii Horst, and **c**-ii Tilted fault blocks

Fault blocks are the fingerprints of extension; each one tells the story of how the crust yields, bends, and breaks under tectonic stress.

6.3.4 Conjugate Faults

Conjugate faults are characterised by two steeply dipping faults with opposing dip directions that intersect or transect each other. These patterns result from mechanical interactions between overburden stresses and ductile layers (e.g., evaporites) (Baudon & Cartwright, 2008; Needham et al., 1996).

During gravity-driven tectonics, evaporite movement interacts dynamically with overlying sediment loading, generating stresses that promote conjugate geometries. In some cases, conjugate geometries also develop through the interaction of planar

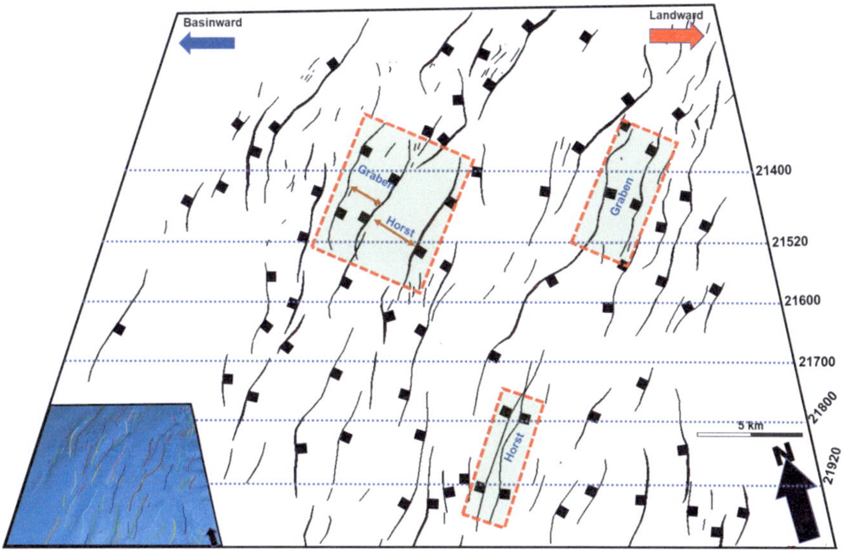

Fig. 6.8 Map view of fault polygon showing block faulting as graben–horst systems across the Levantine Margin. Oppositely dipping normal faults define downthrown grabens and uplifted horsts, creating alternating subsiding and elevated fault blocks within NE–SW–trending fault arrays. Red dashed outlines highlight representative graben and horst structures, blue dashed lines indicate regional seismic profiles (Lines 21,400–21,700), and arrows show basinward and landward directions. The diagram illustrates how block faulting divides extensional strain and organises basin architecture within a gravity-driven rift setting

Table 6.2 Classification of faults based on structural styles, fault scale, fault polarities, and interrelationships of faults identified from the 2D regional seismic profiles

Structural styles		Lines					
		21,920	21,820	21,700	21,600	21,520	21,400
	Listric	6	6	7	4	7	6
	Planar	81	51	97	82	70	98
Conjugate faults	V-conjugate	P	P	P	P	P	P
	X-conjugate	P	P	A	A	P	P
	Y-conjugate	P	P	P	A	P	P
Block faulting	Graben	P	P	P	P	P	P
	Horst	P	P	P	P	P	P
	Tilted faults	P	P	P	P	P	P

*P represent Present, A represent Absent

and listric faults with opposing dip directions. Three conjugate fault geometries were identified: (i) V-shaped; (ii) Y-shaped; and (iii) X-shaped (see Figs. 6.9, 6.10 and 6.11).

I. *V-shaped Conjugate Faults*

V-shaped conjugate faults consist of two planar faults dipping in opposite directions, with their tips meeting at depth or near the surface, so that their upper or lower tips intersect at a common point. These intersections often occur within the Messinian units, forming fault-bounded graben-like structures (see Figs. 6.9a and 6.10a). V-shaped conjugate faults indicate early linkage between fault segments during symmetric extension.

V-shaped conjugate geometries are frequently observed in the Levantine margin. They occur in all seismic facies and are present on lines 21,920, 21,820, 21,700, and 21,400 (see Table 6.2).

ii. *Y-shaped Conjugate Faults*

Y-shaped conjugate faults form when the tip of one fault intersects and ends against the length of an opposing, dipping fault that is larger in extent. The minor fault intersects the primary fault, causing a clear asymmetry in the conjugate set (e.g., Dawers et al., 1993; Fig. 6.9b and 6.10b). Two main geometries occur: where two opposed dipping planar faults intersect, and where a planar fault intersects an opposed dipping listric fault.

Y-shaped conjugates are widely distributed throughout the Levantine margin. They occur in all seismic facies units but are particularly abundant along the eastern margins of Lines 21,920, 21,820, 21,700, and 21,520 (see Table 6.2). Their prominence at margin-scale locations suggests that Y-shaped conjugates accommodate significant lateral strain at the termination of fault propagation.

iii. *X-shaped Conjugate Faults*

X-shaped conjugate faults occur when two planar faults with opposing dips intersect, forming a symmetrical "X" structure (e.g., Dawers et al., 1993; see Fig. 6.9c and 6.10c). Unlike Y-shaped sets, in which a minor fault terminates in a major one, X-shaped conjugates form true intersections with no clear hierarchy.

X-shaped conjugates are present in both the Messinian units and the Plio-Quaternary strata. Their occurrence is particularly prominent on lines 21,920 and 21,820 (see Table 6.2). These geometries reveal zones of intense, localised stress, where faults extend towards each other and intersect, often forming complex relay or linkage structures.

In map view, conjugate faults are recognised as overlapping or collateral faults (see Fig. 6.11). These conjugate systems often indicate linked fault architectures (Nicols et al., 1995; Higgins, 2007). This variability demonstrates the dynamic nature of fault propagation, where lateral shifts in displacement direction produce distinct expressions of extensional deformation. The conceptual model of the conjugate faults is shown in Fig. 6.12b.

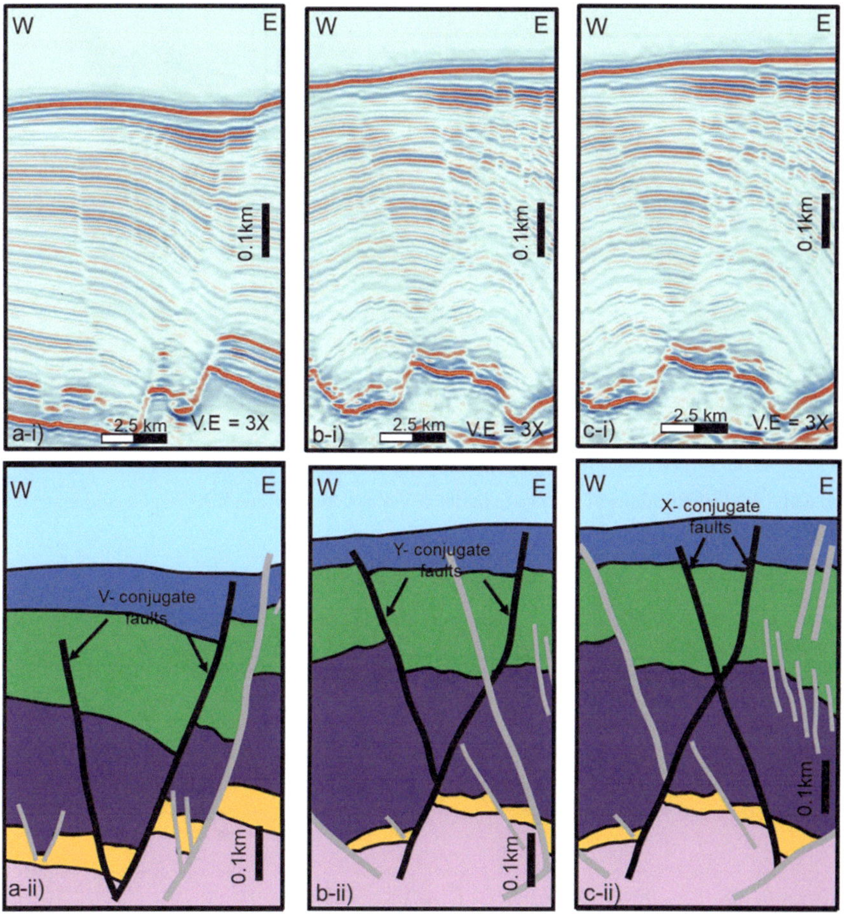

Fig. 6.9 Seismic sections illustrating conjugate fault geometries. Sections (**a**-i), (**b**-i), and (**c**-i) depict uninterpreted seismic profiles. Geoseismic sections showing the interpreted faults, (**a**-ii) V-conjugate faults, (**b**-ii) Y-conjugate faults, and (**c**-ii) X-conjugate faults

On 2D seismic profiles, conjugate faults appear as V-, X-, and Y-shaped features. On the map, conjugate faults in the Levantine Margin exhibit overlapping or collateral geometries. Their displacement directions determine whether they form graben-like convergent or horst-like divergent systems.

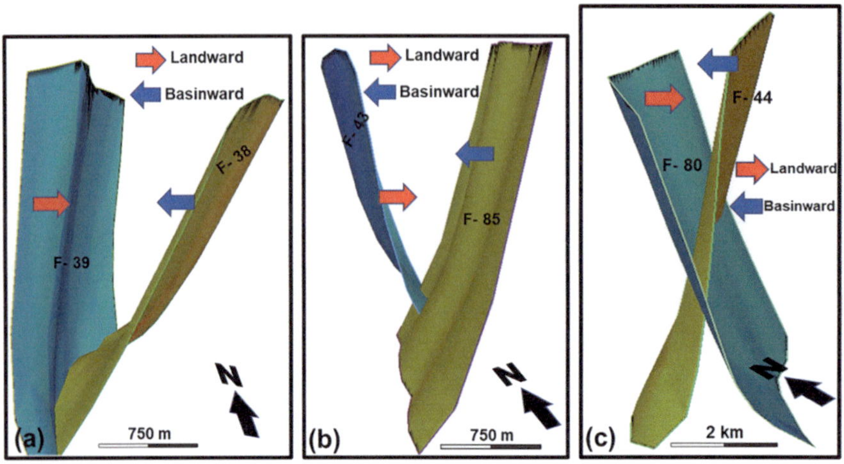

Fig. 6.10 3D visualisations of conjugate faults: **a** V-shaped conjugate, **b** X-shaped conjugate, and **c** Y-shaped conjugate

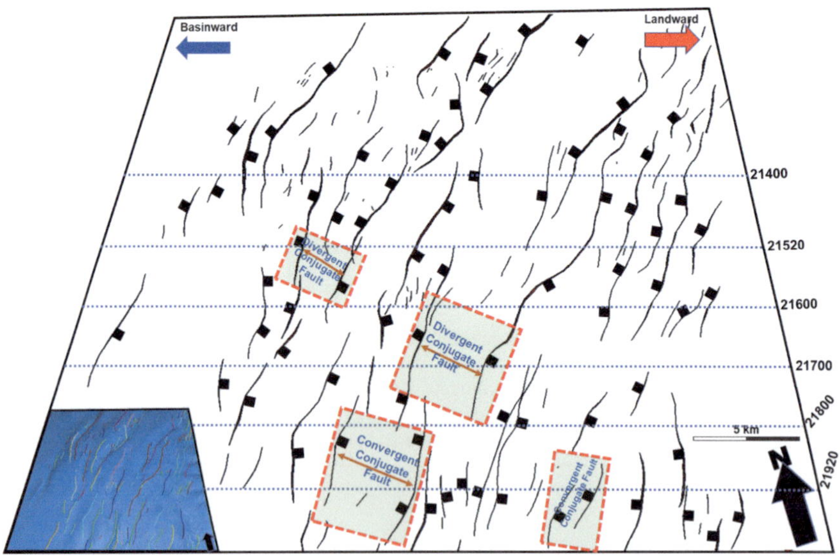

Fig. 6.11 Map view of fault polygons illustrating conjugate fault systems across the Levantine Margin. Oppositely dipping normal faults intersect to form convergent and divergent conjugate geometries, expressed as V-, Y-, and X-type interactions on the map. Convergent conjugates produce graben-like structures, while divergent conjugates generate horst-like geometries, reflecting local variations in displacement direction and stress interaction. Red dashed envelopes highlight representative conjugate fault sets, blue dashed lines mark regional seismic profiles (Lines 21,400–21,700), and arrows indicate basinward and landward directions. The diagram emphasises how conjugate faulting accommodates extensional strain through fault interaction and linkage within a gravity-driven rift system

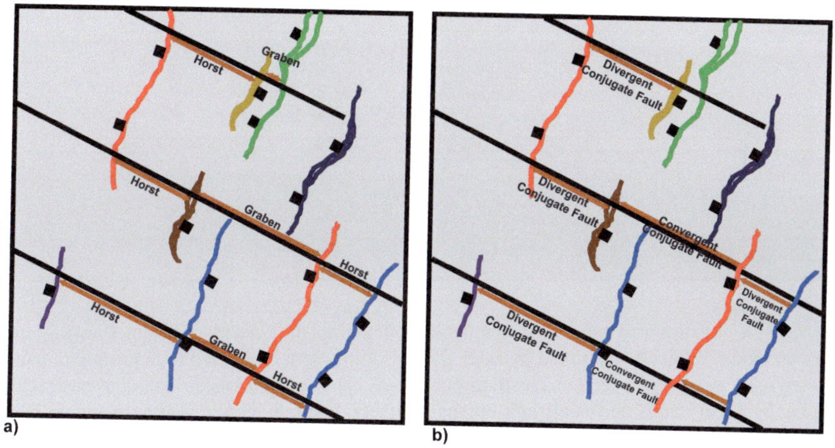

Fig. 6.12 Conceptual Map View Model of: **a** Block Faulting (e.g., Graben and Horst); **b** Conjugate Faults (e.g., Convergent and Divergent)

6.4 Fault Polarities

Fault polarities indicate the direction of fault displacement in an extensional regime. These faults dip either east or west, extend into the deeper basin, and are part of the gravity-driven collapse system. Each polarity contributes to the total displacement (throw or heave), which is distributed between faults dipping in opposite directions. The sum of these displacements determines the structural equilibrium of the margin. Two main fault polarities are recognised: (i) synthetic faults (dipping basinward) and (ii) antithetic faults (dipping landward).

6.4.1 Synthetic Faults

Synthetic faults are fault planes that display the same shear sense of displacement as the master fault or principal deformation zone (see Fig. 6.13). Unlike antithetic faults, which oppose the motion of the main structure, synthetic faults act as companions, accommodating displacement in parallel with their associated fault.

From the regional 2D seismic profiles, synthetic faults are identified within both the Messinian units and the Plio-Quaternary strata, with fault lengths varying across the margin. They occur along both the eastern and western margins of Lines 21,920, 21,820, 21,700, 21,600, 21,520, and 21,400 (see Table 6.3). Quantitative analysis shows alternating fault abundances from west to east across the regional sections: the highest number of synthetic faults is on Line 21,520, with 35 faults, while the lowest is on Line 21,820, with 19 faults (see Table 6.3; Fig. 6.14).

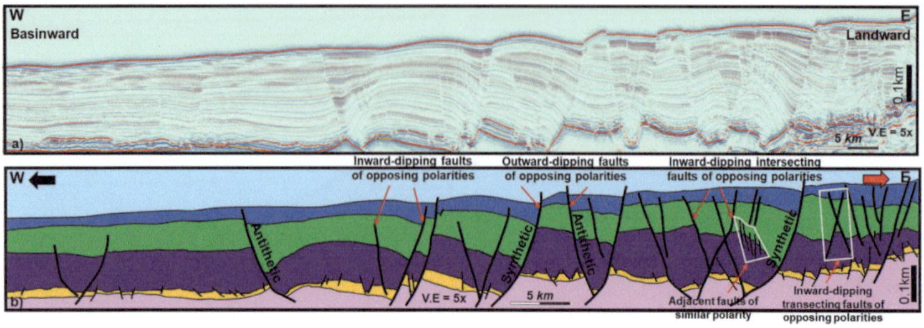

Fig. 6.13 **a** Uninterpreted regional W–E seismic profile across the Levantine Margin, showing laterally continuous reflectors disrupted by extensional faulting from basinward to landward domains. **b** Interpreted geoseismic section illustrating faults of opposing polarities, including basinward-dipping synthetic faults and landward-dipping antithetic faults, transecting Messinian and overlying Plio–Quaternary units. Adjacent and intersecting fault sets form graben–horst geometries and locally convergent and divergent fault arrangements, highlighting asymmetric strain partitioning within a gravity-driven extensional system. Vertical exaggeration (V.E.) = 5×; horizontal scale = 5 km; vertical scale = 0.1 km

Table 6.3 Statistical distribution of synthetic and antithetic faults across regional seismic lines

Line	Synthetic faults (Count)	Antithetic faults (count)	Remarks
21,920	37	50	Both polarities are present on the eastern and western margins
21,820	19 (lowest)	38 (lowest)	Structural low point: both polarities exhibit the lowest counts
21,700	32	72 (highest)	Antithetic maximum indicates asymmetric strain accommodation
21,600	32	54	Both polarities present; values between extremes
21,520	35	42	Balanced distribution of polarities
21,400	47 (highest)	57	The synthetic maximum reflects the basinward dominance of extension

In the map view, synthetic faults are identified by fault polarities that match the shear sense and orientation of the main fault. Consistent with observations from the 2D regional profiles, synthetic faults appear both as individual structures and as basinward-facing arrays (see Fig. 6.15). Their key characteristic is a parallel dip direction that corresponds with the regional extensional stress field, confirming their role as subordinate yet cooperative parts of the overall fault system.

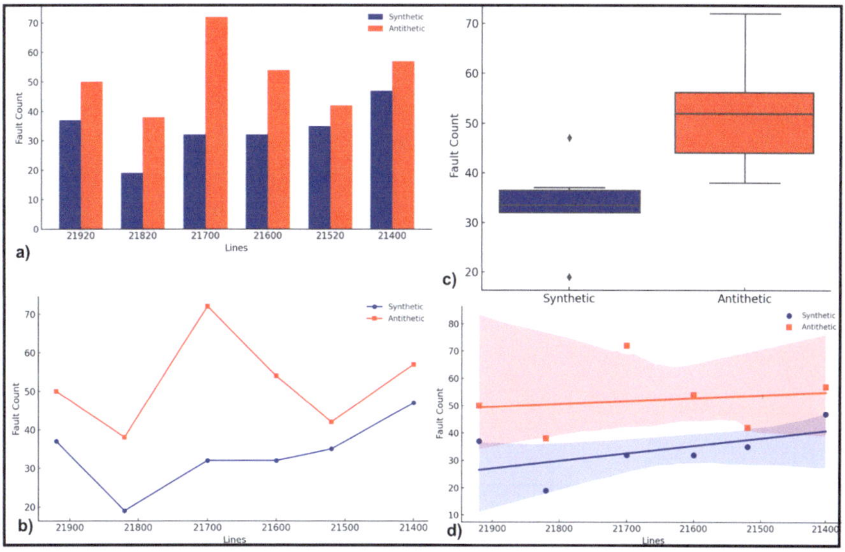

Fig. 6.14 Shows the distribution of synthetic (blue) and antithetic (red) faults, with regional 2D seismic profiles (Lines 21,920, 21,820, 21,700, 21,600, 21,520, and 21,400). **a** A bar chart displays the total fault counts per interpreted seismic line. **b** The spatial variation of synthetic and antithetic fault counts is plotted against seismic line position, revealing polarity-dependent divergence along strike. **c** Box-and-whisker plots summarise fault count distributions for both fault polarities, indicating a higher median and greater variance in the antithetic population. **d** Linear trend envelopes illustrate fault scaling behaviour and count dispersion ranges, supporting a consistent dominance of antithetic faults, with peak divergence seen on Line 21,700. The results suggest along-strike structural segmentation and variable fault-interaction strength, influenced by fault polarity

6.4.2 *Antithetic Faults*

Antithetic faults are secondary faults with displacement directions opposite to that of the main fault or deformation zone (see Fig. 6.13). As they often branch from dominant structures, they are commonly called branched faults.

Antithetic faults are identified in regional 2D seismic profiles within both the Messinian units and the Plio-Quaternary strata, displaying varying lengths. They are widespread across the margin, occurring on both the eastern and western sectors of Lines 21,920, 21,820, 21,700, 21,600, 21,520, and 21,400 (see Table 6.3). A quantitative assessment shows an alternating pattern of fault abundance from west to east along the margin. The highest number of antithetic faults is on Line 21,700, with 72, while the lowest is on Line 21,820, with 38 (see Table 6.3 and Fig. 6.14).

In map view, antithetic faults display fault polarities, displacement directions, and orientations that are opposite to those of the dominant fault system (see Fig. 6.15). Unlike synthetic faults, which follow the main shear sense, antithetic faults dip in the opposite direction of regional extension, making them the structural counterparts of synthetic faults. As shown in the 2D regional profiles, antithetic faults on fault maps

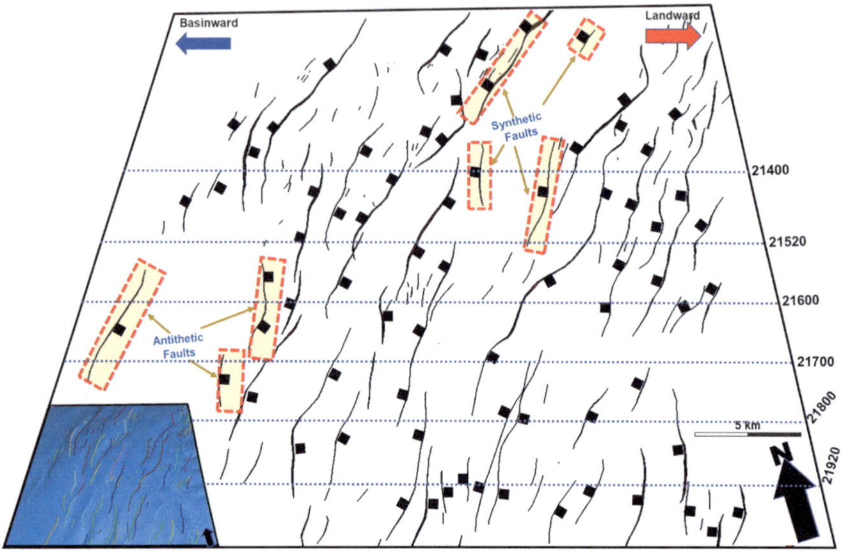

Fig. 6.15 Map view of fault polygons illustrating fault polarity patterns across the Levantine Margin. Basinward-dipping synthetic faults and landward-dipping antithetic faults coexist within NE–SW–trending fault arrays, accommodating extension through opposing displacement directions. Red dashed envelopes highlight representative synthetic and antithetic fault segments, blue dashed lines mark regional seismic profiles (Lines 21,400–21,700), and arrows indicate basinward and landward orientations. The diagram demonstrates how variations in fault polarity reflect asymmetric strain partitioning and spatial reorganisation of extensional deformation along the margin

appear as both outward- and inward-dipping structures, but there is no systematic relationship between them.

The conceptualised model of fault polarities shows their fault patterns, displacement directions, and orientations (see Fig. 6.16).

Synthetic faults in the Levantine Margin show the master fault shifting basinward due to regional extension, appearing as isolated structures and arrays. Antithetic faults oppose shear, dipping in the opposite direction to extension, especially in southern basinward zones, forming grabens and horsts with synthetic faults.

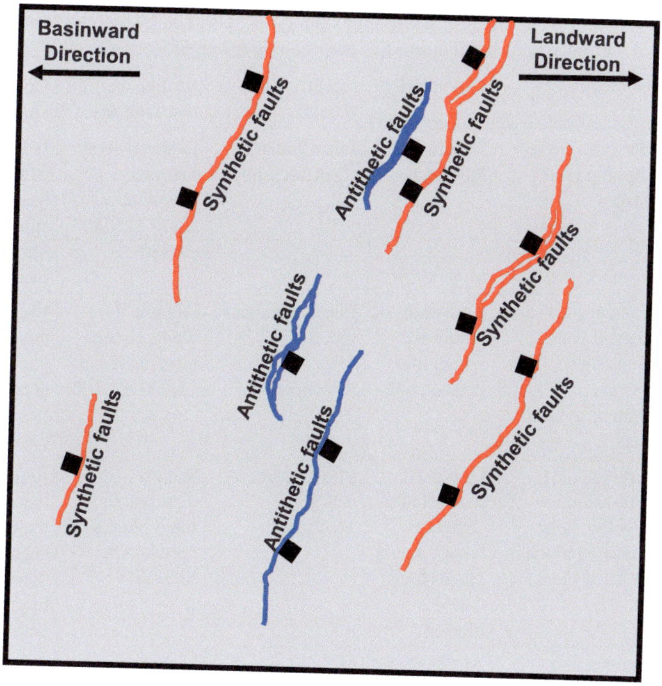

Fig. 6.16 Conceptualised Map view model of fault polarities (e.g. Antithetic and Synthetic)

6.4.3 Statistical Analysis of Fault Polarities

Analysis of the regional 2D seismic profiles confirms the presence of both synthetic and antithetic faults across multiple lines, including Lines 21,920, 21,820, 21,700, 21,600, 21,520, and 21,400 (see Table 6.3). The counts indicate apparent lateral variations between the western and eastern margins of the profiles.

The number of synthetic faults varies. The highest count is at Line 21,400, with 42 faults, while the lowest is at Line 21,820, with 19 faults (see Table 6.3; Fig. 6.14). These differences indicate that synthetic faults tend to cluster in areas where extension strongly aligns with the main fault trend.

Antithetic faults consistently surpass synthetic faults across all seismic lines, with the largest difference observed on Line 21,700, where fault interaction and segmentation are most pronounced. Line 21,700 has 72 faults, while Line 21,820 has the fewest, with 38 faults (see Table 6.3 and Fig. 6.14). This pattern highlights the asymmetric distribution of strain, with antithetic geometries dominating some areas but remaining subordinate in others.

Comparative summaries of structural styles indicate various classifications based on geometric characteristics, spatial occurrence and distribution, and geological significance (see Table 6.4).

Table 6.4 Comparative summary of interpreted fault types classified by their geometric characteristics, spatial occurrence and distribution, and their geological significance

Structural style	Geometry	Dip/polarity	Occurrence (Units)	Spatial distribution	Geological significance
Planar faults	Straight, steeply dipping fault planes	Basinward or landward	Plio-Quaternary (Post-Messinian)	East and west margins; abundant in Lines 21,700, 21,920	Indicate early rifting; control basin subsidence and sediment pathways
Listric faults	Concave-upward, curved planes; steep near surface, flattening with depth	Basinward or landward; merge into detachments	Plio-Quaternary; link to Messinian detachments	Dominant along eastern margins (Lines 21,920–21,400)	Create rollover anticlines; generate accommodation space; control depocentres
Block faulting	Graben (subsiding blocks), horst (uplifted blocks), or tilted blocks	Opposed planar faults (graben/horst); single planar faults (tilted blocks)	Plio-Quaternary and Top Messinian	Widely distributed; tilted blocks common in SP2–SP3	Reflect extensional tectonics; segment basins; create traps and migration pathways
Conjugate faults	Intersecting faults of opposite polarity (V-, Y-, or X-shaped)	Antithetic vs synthetic dip combinations	Messinian and Plio-Quaternary	V-shaped: margin-wide; Y-shaped: eastern margins; X-shaped: localised intersections	Record stress interactions; form complex relay/linkage zones; influence basin architecture
Antithetic faults	Subsidiary faults opposing master displacement ("branched faults")	Landward	Messinian and Plio-Quaternary	Both margins; highest abundance at Line 21,700	Accommodate strain; influence local trapping and the balance of the fault array
Synthetic faults	Subsidiary faults paralleling master displacement	Basinward	Messinian and Plio-Quaternary	Both margins; highest abundance at Line 21,520	Reinforce master fault kinematics; balance displacement in extensional arrays

Statistical analysis indicates that fault polarities vary across the Levantine Margin, with synthetic faults peaking at Line 21400 (42 faults) and antithetic faults at Line 21700 (72 faults). Both fault types reach their lowest on Line

21820, suggesting that differential stress and detachment geometry influence fault polarity.

6.5 Fault Attribute Analysis Using Map Visualisations

Fault attribute maps provide quantitative insights into fault displacement, orientation, and mechanical behaviour. These attributes reveal variations in strain distribution, fault maturity, and structural complexity. The three main attribute maps created are fault-heave distribution, dip azimuth, and dip angle. These parameters compare structural interpretation, fault growth analysis, and evaluation of mechanical stratigraphy (see Table 6.5).

6.5.1 Fault Heave Distribution Map

The fault-heave distribution map shows lateral displacement along fault planes, with values ranging from 0 to 800 m. Areas with higher heave indicate zones of significant

Table 6.5 Comparative summary of fault attribute maps of the levantine margin

Fault attribute	Definition	Range	Tectonic significance	Insights from the levantine margin
Heave distribution	Lateral displacement is accommodated along fault planes	0–800 m	Indicates zones of strain accumulation and horizontal slip; highlights active deformation domains	High values of heave mark tectonically active zones; variations reveal differential stress and fault growth
Dip azimuth	Compass direction of fault plane inclination	0°–360°	Defines dominant fault orientations; links fault geometry with regional stress regime	NNE–SSW orientations dominate; variations indicate linkage and relay ramp development
Dip angle	Inclination of the fault plane relative to the horizontal	0°–90°	Differentiates steep planar faults from shallow detachments; controls displacement style and reactivation	Steep faults accommodate vertical displacement; shallow faults merge into detachments, aiding salt-related deformation

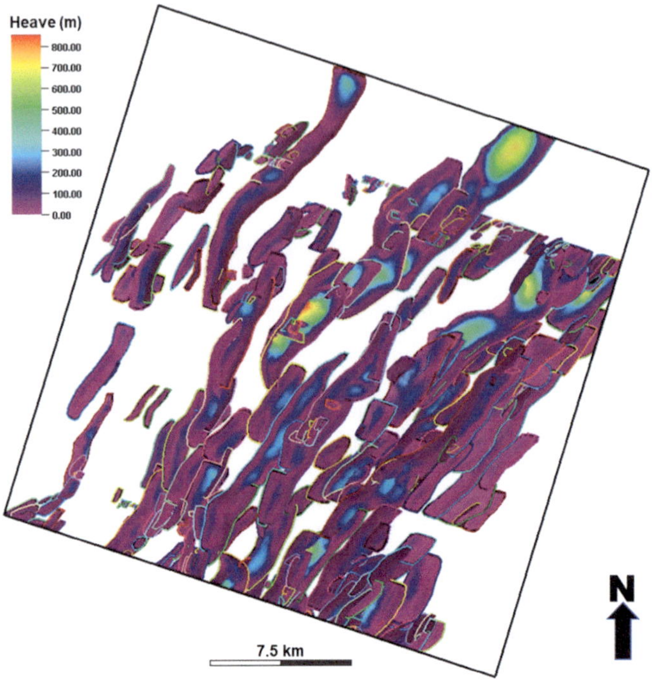

Fig. 6.17 Fault heave distribution map, illustrating variations in lateral fault displacement, with values ranging from 0 to 800 m

horizontal movement and increased strain accommodation (see Fig. 6.17). Heave is affected by fault dip, curvature, mechanical layering, and displacement gradients along strike. Higher heave values often occur near large planar faults (Gibbs, 1984), steep-dipping normal faults, and segments that link or overlap (Dawers et al., 1993; Needham et al., 1996).

The fault attribute map shows that apparent planar faults have heave values ranging from 0 to 150 m, whereas apparent listric faults exceed 150 m (see Fig. 6.17). This aligns with traditional geometric and mechanical principles of normal fault systems (e.g., Fossen & Rornes, 1996; Needham et al., 1996). Variations in heave and dip angle with depth reflect structural complexity and mechanical heterogeneity.

6.5.2 *Dip Azimuth Map*

The dip azimuth map shows the orientation of fault planes, with azimuth values ranging from 0° to 360° (see Fig. 6.18). Dominant trends include NE–SW striking planar faults, NNE–SSW striking listric faults and opposing dips in conjugate sets (Childs et al., 1995; Fossen & Rornes, 1996). These orientations indicate the compass

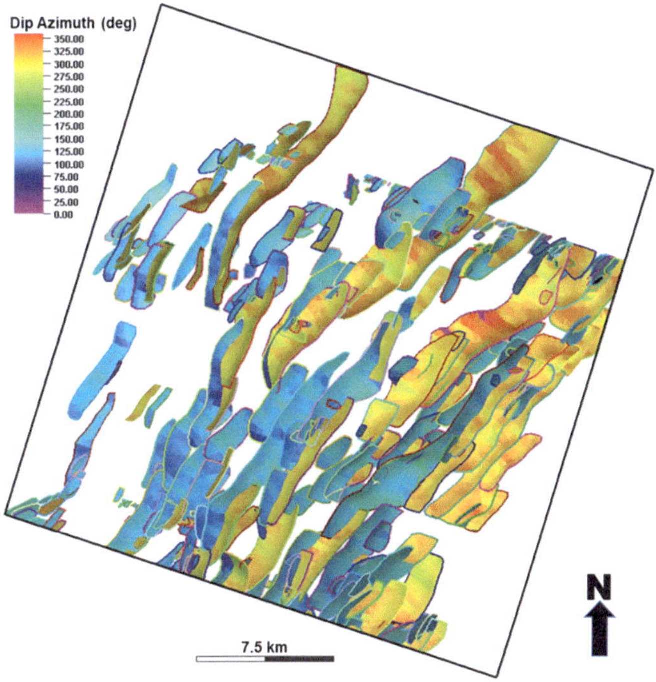

Fig. 6.18 Dip azimuth map displaying the orientation of fault planes, with azimuth angles from 0° to 360°, indicating primary fault trends

direction of dip, allowing identification of the main fault trends. By correlating azimuth patterns with regional stress fields, we can understand how individual fault segments integrate into larger fault networks. This is essential for linking fault kinematics to tectonic forces and for recognising whether faults formed synchronously or sequentially under changing stress regimes.

The dip-azimuth map, known as the "polarity map," segmented by dip-azimuth and fault type, shows clear distribution patterns. Antithetic faults (blue) are more common and spread over a wider area, especially in the southern, basinward region. In contrast, synthetic faults (yellow) are more prevalent landward but less dense. This polarity imbalance suggests a structurally complex rift basin characterised by asymmetric extension and distributed deformation.

6.5.3 *Dip Angle Map*

The dip-angle map displays the inclination of fault planes, ranging from 0° to 90° (Fig. 6.19). Variations in dip angle highlight differences in fault steepness across the Levantine Margin. Steeper dip angles (>60°) are associated with steep planar fault

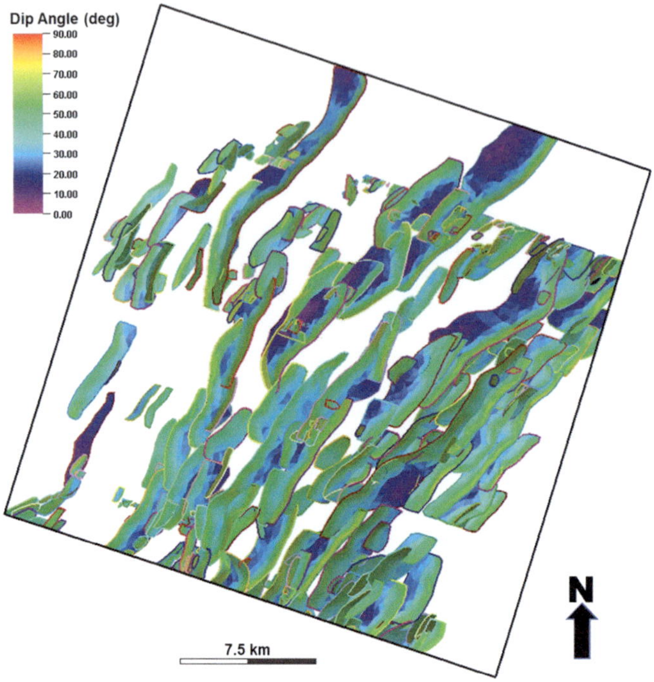

Fig. 6.19 Dip-angle map displaying fault-plane inclinations from 0° to 90°, illustrating variations in fault steepness

segments, dominant dip-slip displacement, and extensional tectonic regimes (e.g., Lowell, 1985). Lower dip angles (<30°) suggest more horizontal movement, linked to listric flattening, mechanical layering effects, and detachment-level adjustments (e.g., evaporites) (e.g., Baudon & Cartwright, 2008). These differences reflect diverse mechanical behaviours across fault populations.

The combined attribute maps show:

i. Zones of high heave and steep dip angles align with major fault corridors.
ii. Azimuth variability indicates multiphase fault development with rotations.
iii. Shallow dip angles suggest listric detachment levels and potential salt-related deformation (e.g. Baudon & Cartwright, 2008).

These observations align with segmentation and linkage behaviour (Childs et al., 1990; Needham et al., 1996), indicating different fault maturity levels from FM-1 to FM-4 and variations in the distribution and localisation of strain accommodation. Fault attribute correlations support subsequent quantitative analysis of fault scaling relationships.

Fault attribute maps serve as powerful visual tools for analysing structural geometry. Heave distribution maps emphasise lateral strain partitioning, dip azimuth maps showcase the orientation of regional stress regimes, and dip angle maps differentiate fault steepness and reactivation potential.

6.6 Conclusion

This chapter synthesises fault geometry, structural style, polarity, and attribute analyses to establish a consistent framework for understanding the extensional architecture of the Levantine continental margin. Integrated interpretation of regional 2D seismic profiles and high-resolution 3D fault maps shows that closely spaced normal faults organised into systematic NE–SW to NNE–SSW arrays dominate both the vertical and lateral structure of the basin. Fault length populations reveal a hierarchical organisation, where very small and small faults record early-stage strain partitioning, medium faults display progressive growth and interaction, and long faults define rift-scale architecture and spacing governed by mechanical stratigraphy and detachment geometry.

Different structural styles, like apparent planar and listric faults, block faulting, and conjugate fault systems, coexist and interact across Messinian and Plio–Quaternary units. Planar faults dominate shallow layers and form linear fault arrays, while listric faults handle extension through curvature, rollover, and linkage to detachment horizons. Graben–horst systems and tilted blocks show distributed and rotational extension. V-, Y-, and X-shaped conjugate faults mark stress zones, fault linkage, and localised strain.

Fault polarity analysis shows more antithetic than synthetic faults along the margin, indicating asymmetric extension and strain variability. Differences in polarity, fault abundance, and displacement highlight the roles of sediment loading, salt movement, and detachment geometry on fault development. Attribute maps of heave, dip azimuth, and dip angle offer quantitative insights into fault maturity, displacement, and kinematic coherence, linking fault geometries to basin stress regimes and tectonic activity.

The chapter shows that combining geometric classification with quantitative fault attributes provides a strong method for understanding the development of extensional margins. Readers are encouraged to apply this combined workflow to other rifted margins, using fault geometry, orientation, polarity, and attributes as diagnostic tools to interpret tectonic history, strain partitioning, and basin evolution.

Learning Objectives:

After completing this chapter, readers will be able to:

1. Classify fault geometries, including planar, listric, and block-bounding faults.

2. Interpret graben–horst systems and fault polarity patterns.
3. Relate observed fault architectures to extensional tectonic regimes.
4. Analyse how detachment horizons influence thin-skinned deformation.
5. Compare structural styles within the Levantine margin to other rifted margins.

References

Baudon, C., & Cartwright, J. A. (2008). 3D seismic characterisation of an array of blind normal faults in the Levant Basin, Eastern Mediterranean. *Journal of Structural Geology, 30*, 746–760.

Childs, C., Watterson, J., & Walsh, J. J. (1995). Fault overlap zones within developing normal fault systems. *Journal of the Geological Society, 152*, 535–549.

Dawers, N. H., Anders, M. H., & Scholz, C. H. (1993). Growth of normal faults; displacement-length scaling. *Geology, 21*, 1107–1110.

Fossen, H., & Rornes, A. (1996). Properties of fault populations in the Gullfaks Field, northern North Sea. *Journal of Structural Geology, 18*(23), 179–190.

Garfunkel, Z., & Almagor, G. (1985). Geology and structure of the continental Margin off northern Israel and the adjacent part of the Levantine Basin. *Marine Geology, 62*, 105–131.

Gibbs, A. D. (1984). Structural evolution of extensional basin margins. *Journal of the Geological Society, London, 141*(4), 609–620.

Higgins, S. (2007). *3D Seismic analysis of the geometry and development of a deep-water fold and thrust belt*. PhD Thesis, Cardiff University, Chapter 2, 1–18.

Lowell, J. D. (1985). *Structural styles in petroleum geology*. Oil and Gas Consultants International Inc.

Needham, D. T., Yielding, G., & Fox, K. (1996). Fault population description and analysis: Examples from the UK North Sea. *Journal of Structural Geology, 18*(2–3), 159–174.

Nicol, A., Walsh, J. J., Watterson, J., & Bretan, P. G. (1995). Three-dimensional geometry and growth of conjugate normal faults. *Journal of Structural Geology, 17*, 842–862.

Nwosu, O. (2013). *The compartmentalisation of a deepwater fold and thrust belt in the Levant*. PhD thesis, Cardiff University, U.K.

Shelton, J. W. (1984). Listric normal faults: An illustrated summary. *American Association of Petroleum Geologists Bulletin, 68*, 801–815.

Wernicke, B. (1981). Low-angle normal faults in the Basin and Range Province: Nappe tectonics in an extending orogen. *Nature, 291*, 645–648.

Woods, E. (1988). Extensional structures of the Jabiru Terrace. In P. G. Purcell, & R. R. Purcell (Eds.), *The North West Shelf, Australia. Proc Explora. Soc. Australia Symposium*, Perth, pp. 311–330.

Chapter 7
3D Structural Interpretation of Fault Interactions

Abstract This chapter explores how extensional faults interact, connect, and develop into three-dimensional networks rather than existing as isolated structures. Using high-resolution 3D seismic interpretation of the Levantine Margin, it shows that fault growth is a continuous, interaction-driven process influenced by geometry, kinematics, and strain distribution. The analysis combines plan-view fault maps from various stratigraphic levels to illustrate how segmented normal faults overlap, approach, abut, splay, and cross-cut over time. Focus is given to relay zones, where displacement is transferred between neighbouring faults, and on the transition from unbreached relays to single- and double-tip breached linkages. Kinematic relationships are assessed by comparing fault–fault intersection lines with displacement directions, revealing simple, complex, and mixed linkage models that demonstrate different levels of mechanical efficiency. The chapter encourages readers to consider: How does displacement transfer influence fault connectivity? When does strain localise, and why does the style of linkage vary with depth? By combining geometric classification with kinematic and strain-based analyses, it provides a clear framework for understanding fault-interaction mechanisms and reconstructing the evolution of fault networks on extensional continental margins.

Keywords Fault kinematics · Displacement transfer · Segmented faults · Strain localisation · Fault interaction · Fault linkage · Relay ramps

7.1 Introduction

Extensional faults may develop as isolated structures within rifted continental margins (e.g., Kim & Sanderson, 2005; Nicol et al., 2005). In such cases, individual faults nucleate independently and grow by progressive propagation, expressed as radial, step-like, or transtensional advance (e.g., Kim & Sanderson, 2005; Nicol et al., 2005; Peacock, 2002; Walsh et al., 2003). As propagation proceeds, fault length and displacement increase, producing discrete structures that evolve essentially without mechanical interaction with neighbouring faults (e.g., Kim & Sanderson, 2005; Nicol

K. Chiadikobi, *Advances in Horizon Mapping and Fault Interpretation*,
https://doi.org/10.1007/978-3-032-23917-4_7

et al., 2005; Peacock, 2002; Peacock & Sanderson, 1991; Walsh & Watterson, 1989; Walsh et al., 2003).

In many extensional basins, however, faults do not evolve in isolation. Instead, like-dipping faults grow as segmented arrays that overlap along strike and transfer displacement through relay ramps (e.g. Fossen & Rotevatn, 2016; Larsen, 1988; Peacock & Sanderson, 1994). These relay zones accommodate strain between adjacent segments and exert a first-order control on fault linkage, displacement gradients, and hanging-wall deformation. As strain continues to accumulate, relay ramps may be preserved, breached, or transformed into through-going fault surfaces.

Segmented fault systems also branch and intersect, producing conjugate fault relationships (e.g., Odonne and Massonnat, 1992; Kelly et al., 1998; Nixon et al., 2011; Peacock et al., 2017). These interactions commonly produce abutting and cross-cutting geometries that reflect local stress perturbations around propagating fault tips. The resulting structures are not random; instead, they form coherent fault networks in which individual segments share compatible orientations, kinematics, and growth histories imposed by a single regional stress field (e.g., Bull et al., 2006; Fossen & Rotevatn, 2016; Gawthorpe et al., 2003; Peacock & Sanderson, 1991; Peacock et al., 2017). Faults interact mechanically as they propagate, overlap, and connect, continually modifying displacement distributions, strain localisation, and basin architecture.

Most faults identified in the L form interconnected networks rather than isolated structures. Within these networks, faults vary in length, size, orientation, and spacing, and their spatial proximity promotes interaction and linkage (see Fig. 7.1). Similar fault architectures have been documented in extensional tectonic settings, where fault growth is governed by interactions within evolving fault populations (e.g., Larsen, 1988; Muraoka & Kamata, 1983; Peacock & Sanderson, 1994).

This chapter quantifies fault interactions and linkages in the extensional domain, using map-view analysis of the Plio–Quaternary and Top Messinian units interpreted from 3D seismic data. The analysis evaluates how faults interact within a network by integrating three relationships:

1. Geometry: the spatial arrangement of faults, including their overlap, intersection style, and linkage patterns.
2. Kinematics: displacement distributions along interacting faults and the orientation of displacement directions relative to fault–fault intersection lines.
3. Displacement and strain: the magnitude and sense of slip, and the strain required to accommodate displacement within interaction zones.

By combining these relationships, it provides a coherent framework for interpreting fault interaction mechanisms, linkage evolution, and strain localisation within the extensional fault systems.

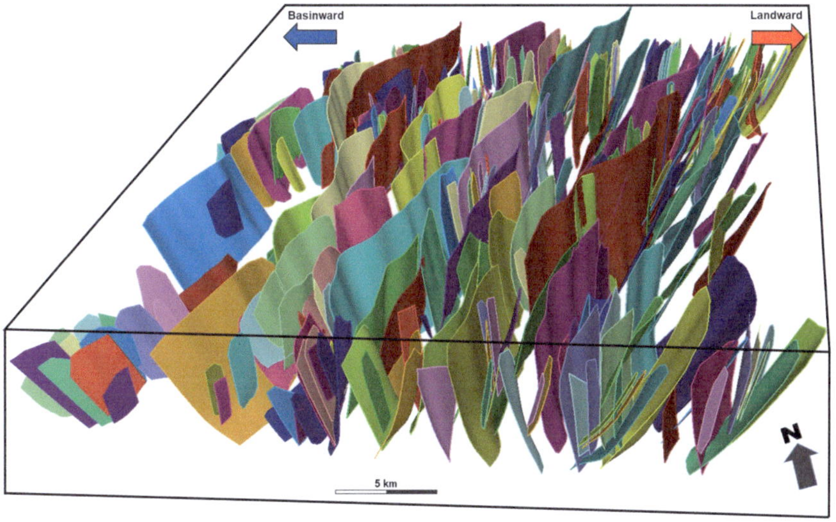

Fig. 7.1 3D view showing the distribution of faults, illustrating clustered fault networks with variations in length, orientation, spacing, and interaction style within the extensional domain

7.2 3D Structural Geometry and Characteristics of Fault Segment Linkages

On the Levantine Margin, a 3D view of fault arrays delineates a series of transfer zones visible across multiple stratigraphic levels (FM-1, FM-2, FM-3, and FM-4; see Figs. 7.2, 7.3, 7.4 and 7.5). Each map captures a distinct stratigraphic window, and together they reveal how fault geometries and interactions evolve with depth and over time.

Fault interaction arises when displacement is transferred between neighbouring faults or when the activity of one fault alters displacement patterns and horizon geometries on adjacent structures. These processes reflect stress redistribution during deformation and control how strain is partitioned within the faulted volume. Variations in throw, fault-tip geometry, and local folding mark zones of interaction commonly provide key evidence for progressive fault growth.

Within these transfer zones, seismic facies display fault-plane geometries, distinct structural styles, and undulating relief features, which are interpreted in places as fold-related deformation. Previous studies show that individual fault linkage types (e.g. Higgins, 2007; Nelson, 2013; Nicol et al., 1995) were differentiated using three key criteria:

1. the stratigraphic level at which linkage occurs,
2. the morphology of the lateral fault-tip region, and
3. the style and intensity of strain in adjacent strata in the transfer zone.

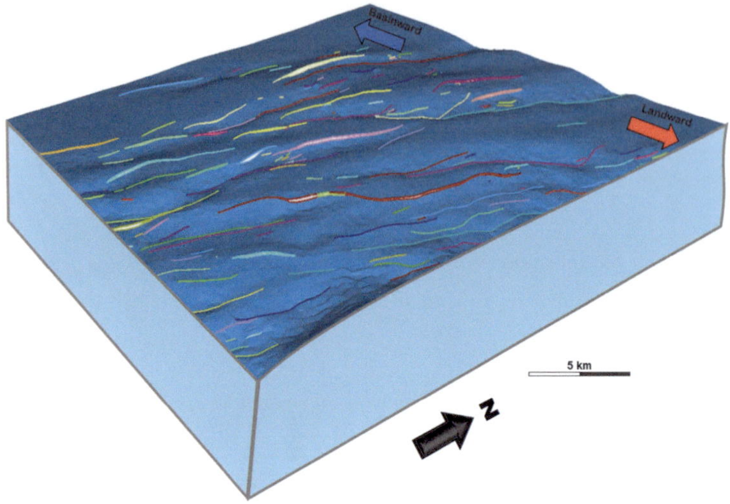

Fig. 7.2 3D perspective view of FM-1 showing the various fault interactions and linkages within the SP1 stratigraphic unit. FM-1 shows early-stage transfer zones and limited fault linkage, with broadly distributed strain and minimal relay breaching

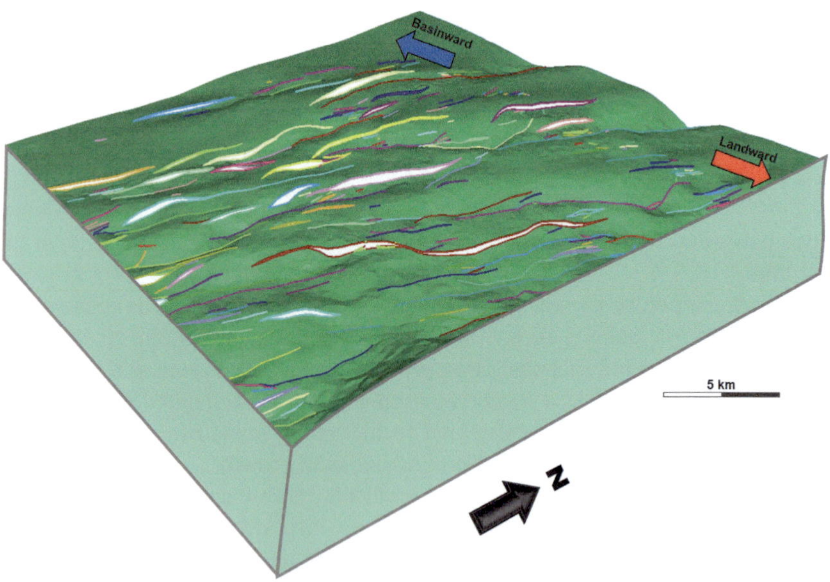

Fig. 7.3 3D perspective view of FM-2 showing the various fault interactions and linkages within the SP1 stratigraphic unit. FM-2 illustrates the development of vertically persistent transfer zones, increased fault overlap, and initial modification of the relay ramp

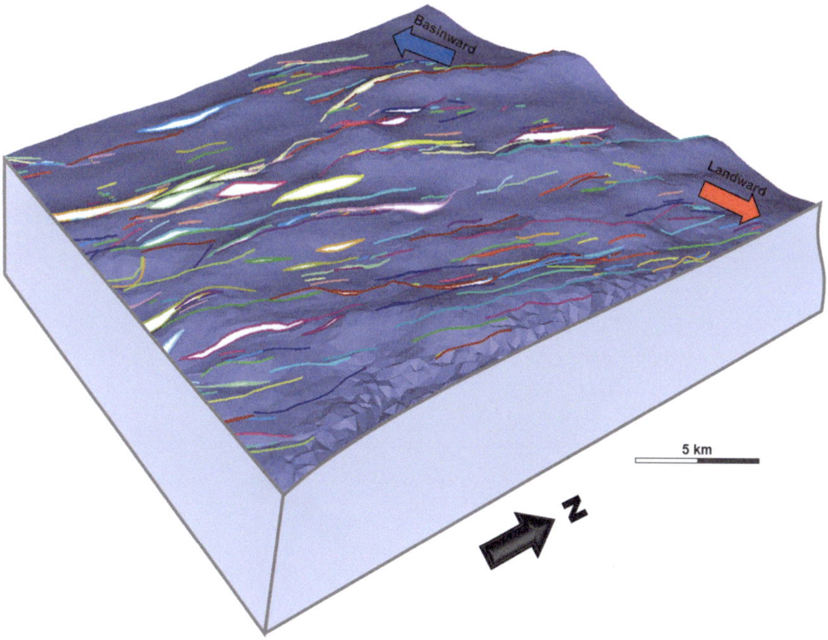

Fig. 7.4 3D perspective view of FM-3 showing the various fault interactions and linkages within the SP3 stratigraphic unit. FM-3 highlights progressive fault linkage, relay ramp breaching, and enhanced branch-line development within the extensional fault network

On this basis, a preliminary classification of transfer-zone architecture is proposed, grounded in the vertical extent of faulting and the geometry of branch lines linking synthetic and antithetic faults. This approach emphasises that upward and downward fault propagation governs linkage style and extensional fault networks.

The interaction and linkage of sub-parallel, synchronously active faults, particularly in extensional regimes, have long attracted focused investigation (e.g., Larsen, 1988; Peacock & Sanderson, 1991; Peacock et al., 2017; Walsh et al., 1999). Such systems rarely evolve as single, through-going structures. Instead, stepping normal faults organise into networks that record successive phases of interaction and segment linkage (e.g., Fossen & Rotevatn, 2016; Peacock, 2002; Peacock & Sanderson, 1991; Peacock et al., 2017).

Within these networks, individual fault segments initiate, propagate, overlap, and interact while remaining active under a common regional stress field. Mechanical interaction between coeval faults promotes displacement transfer and progressive linkage, most commonly between sub-parallel segments as fault tips approach one another, local stress perturbations intensify, guiding fault growth and determining whether linkage occurs through relay ramps, branch lines, or fault breaching.

Extensive analyses of fault displacement patterns and linkage styles, from the initiation of isolated normal faults to the development of overlapping fault systems

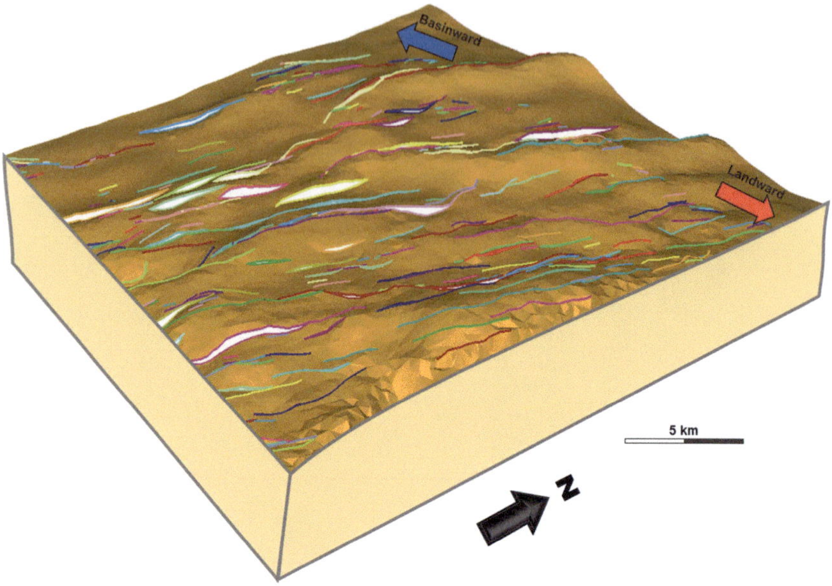

Fig. 7.5 3D perspective view of FM-4, showing the various fault interactions and linkages within the SP4 stratigraphic unit. FM-4 (Top Messinian) shows fully developed transfer zones, complex branch-line geometries, and advanced fault connectivity with pronounced strain localisation

with relay ramps and breached relays, have substantially advanced understanding of extensional fault-system evolution in rift basins. These studies demonstrate that fault growth is progressive and interaction-driven, rather than the product of isolated structures evolving in isolation.

In extensional settings, a single, isolated normal fault provides a valuable reference framework for evaluating fault interaction and linkage behaviour. Against this baseline, neighbouring faults are assessed for overlap, displacement transfer, and geometric compatibility. Faults interact along strike and vertically upward to the Top Messinian, indicating sustained activity over a significant stratigraphic interval.

Most mapped faults cluster along the flanks of fold structures, highlighting a close relationship between faulting and folding. With increasing depth, fault maps show a reduction in fold width across successive stratigraphic levels (FM-1 to FM-4; Figs. 7.2, 7.3, 7.4 and 7.5). This vertical narrowing indicates progressive strain localisation and evolving fault–fold coupling during basin development.

Faults in extensional settings are interconnected, segmented systems influenced by propagation, overlap, stress interactions, and shared tectonic regimes. 3D fault maps show transfer zones are vertically structured with branch-line geometry, horizon offset, and stratigraphic position, which define fault linkage and development. These systems form as coeval, interacting networks with sub-parallel faults linked under a common stress field to accommodate extension. Fault linkage indicates progressive interaction and fold development, best understood through plan-view analysis and 3D visualisation.

7.3 Geometric Relationships of Interacting Faults

Most faults identified in the Levantine Margin do not occur as isolated structures; instead, they form integrated fault networks. Within these networks, faults exhibit a wide range of lengths, sizes, orientations, and spacings, and their spatial proximity promotes interaction and linkage (Fig. 7.1). Such network-style fault architectures are characteristic of extensional systems and have been widely recognised in previous studies, where fault growth and linkage strongly influence basin development and structural complexity (e.g., Larsen, 1988; Muraoka & Kamata, 1983; Peacock & Sanderson, 1994).

Fault interaction occurs when displacement transfers between adjacent faults or when the activity of one fault alters the displacement distribution and horizon geometry of another. These processes show stress redistribution during deformation and illustrate fault system evolution. Zones of interaction often feature changes in throw, fault-tip behaviour, and local folding, indicating ongoing fault growth.

The geometric relationships of fault–fault interactions are defined by the way faults approach, intersect, or terminate against one another. Interactions between pairs of faults can be grouped into four fundamental geometric types: (1) relay faults, (2) approaching faults, (3) abutting and splaying faults, and (4) cross-cutting faults.

7.3.1 Relay Faults

Relay faults describe a fault interaction pattern in which displacement is transferred from one fault to another without direct intersection (e.g., McClay & Khalil, 1998; Morley et al., 1990; Peacock & Sanderson, 1995; Peacock et al., 2017; Walsh & Watterson, 1991). Instead, the faults overlap, forming a relay zone that accommodates strain. This geometry represents a fundamental mechanism for the transfer of displacement during fault growth and is widely recognised in extensional fault systems.

This chapter demonstrate well-developed relay zones as overlapping fault segments (see Fig. 7.6), classified into four types based on fault dip and overlap geometry.

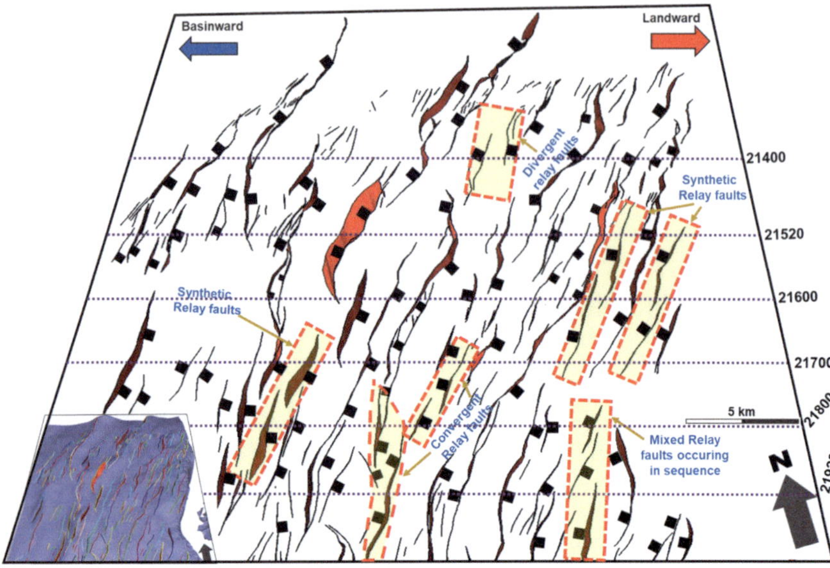

Fig. 7.6 Representative plan view (FM-3) illustrating overlapping relay fault geometries, highlighting convergent, divergent, synthetic and mixed sequences of convergent and divergent relay zones

1. Overlapping convergent zones—formed where two faults overlap and dip towards each other.
2. Overlapping divergent zones—develop where overlapping faults dip away from each other.
3. Overlapping synthetic zones—formed where overlapping faults dip in the same direction.
4. Mixed relay systems—occur where convergent and divergent relay geometries appear in sequence

In the map view, a convergent relay causes displacement vectors to point into the relay zone, thereby concentrating strain. Divergent relay zones produce outward-facing displacement vectors and a widening overlap region. A synthetic relay zone enables kinematically compatible displacement transfer with limited wall-rock deformation. Mixed relay systems occur where convergent and divergent relay geometries appear sequentially along strike, reflecting lateral variations in stress orientation and fault growth history (see Fig. 7.6).

The spatial distribution of these relay zones shows that fault interaction is highly organised rather than random. Relay geometries recur along strike and across the basin, indicating that displacement transfer, strain localisation, and fault linkage are controlled by variations in fault dip polarity, overlap distance, and displacement direction.

The 3D perspective view shows that convergent (see Fig. 7.7a) and divergent (see Fig. 7.7b) relay zones occur preferentially between faults with opposing dip directions. By contrast, synthetic relay zones develop where adjacent faults share the same dip directions (see Fig. 7.7c). A mixed relay system, in which divergent and convergent relay zones occur sequentially along strike (see Fig. 7.7d), is also observed. This configuration highlights the dynamic evolution of relay zones, driven by fault propagation, interaction, and stress reorganisation.

The conceptual model of the overlapping relay fault system is shown in Fig. 7.8. These models demonstrate that overlapping relay faults are transient yet fundamental structures that control displacement continuity, strain localisation, and the transition from isolated faults to linked fault systems.

Relay faults can be further subdivided according to the degree of linkage between overlapping fault segments:

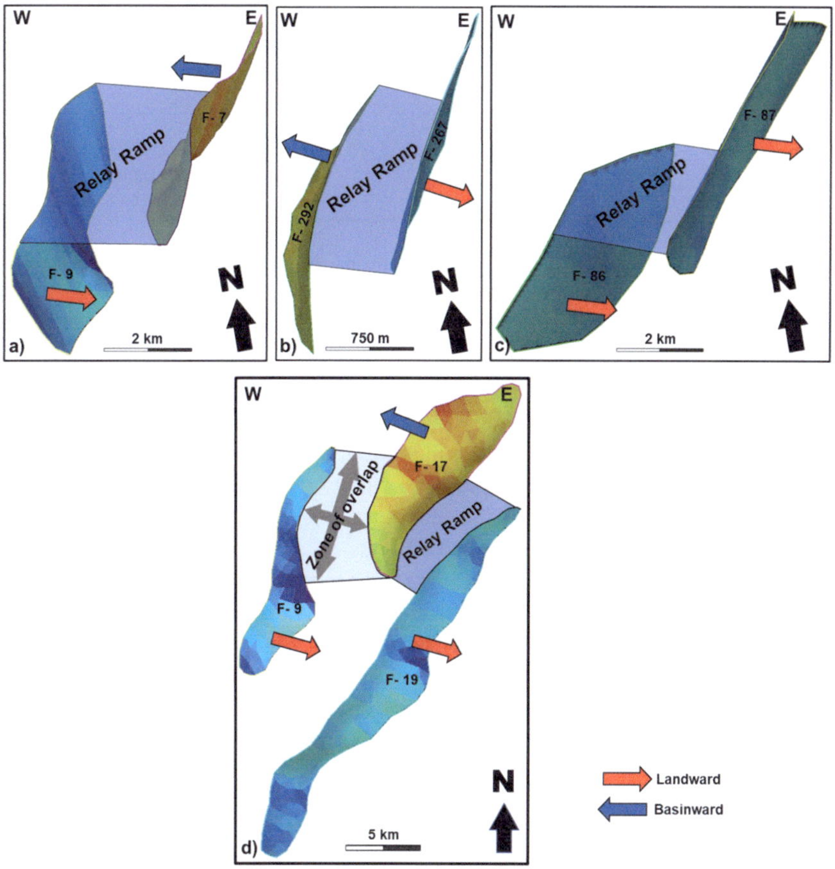

Fig. 7.7 3D perspective view illustrating various types of overlapping relay faults. **a** Convergent relay fault (F-9 and F-7). **b** Divergent relay fault (F-292 and F-267). **c** Synthetic relay fault (F-86 and F-87). **d** Mixed (divergent and convergent) relay fault (F-19, F-17 and F-9) occurring in sequence. The red arrow signifies landward faults, whereas the blue arrow shows basinward faults

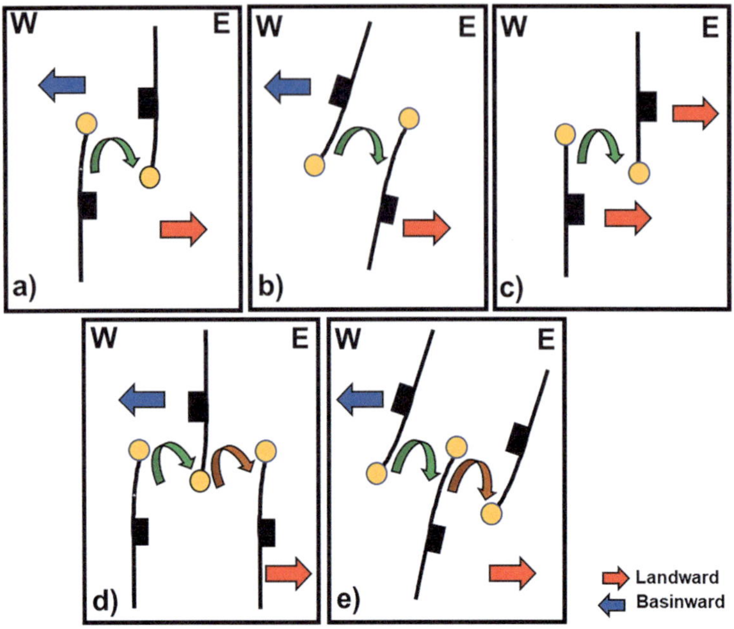

Fig. 7.8 Conceptual model showing various types of overlapping relay faults. **a** Convergent relay faults; **b** Divergent relay faults; **c** Synthetic relay zones; **d** and **e** Mixed (divergent and convergent) overlapping relay faults occurring in various sequences. The arrows indicate the displacement direction of the faults, with basinward in blue and landward in red. The green and brown curved arrows indicate the overlap or transfer zones

1. Unbreached relay faults- In this configuration, two faults overlap without a physical connection. Displacement transfer occurs entirely across the intact relay zone and represents an early stage of fault linkage (see Fig. 7.10a).
2. Single-tip (upper-ramp) breached relay faults - occur when one fault tip becomes arrested while the opposing tip propagates, bends, and ultimately connects with the neighbouring fault. This process reflects asymmetric fault growth and the progressive localisation of strain within the relay zone (see Fig. 7.10b).
3. Double-tip breached relay faults- In this advanced configuration, both overlapping fault tips propagate and link, thereby fully breaching the relay zone and forming a continuous through-going fault (see Fig. 7.10c).

In map view, the connecting fault segment is curved, indicating progressive linkage via relay breaching. The upper (hanging-wall) fault tip propagates across the relay zone, creating an upper-ramp breach. This linkage style is shown in Fig. 7.9, where asymmetric fault growth occurs, with one fault tip transferring most displacement and the other remaining arrested.

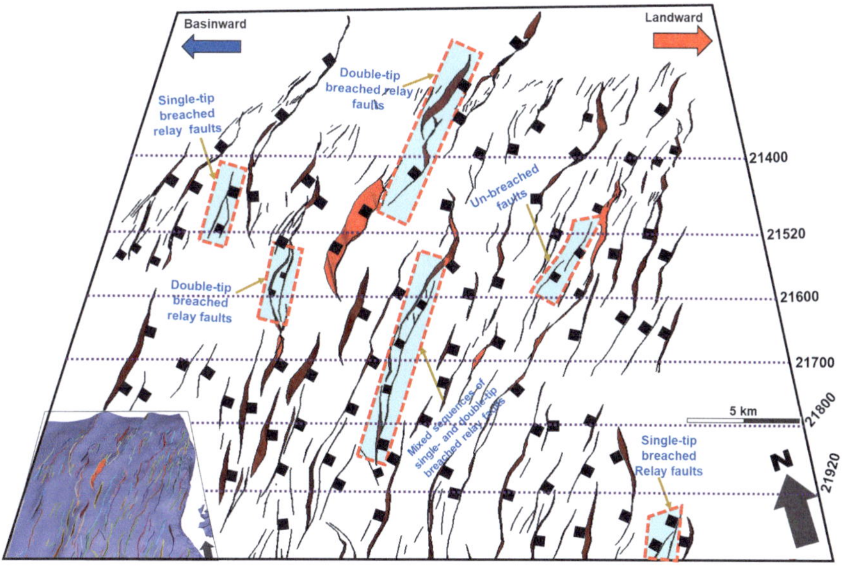

Fig. 7.9 Representative plan view (FM-3) illustrating overlapping and breached relay fault geometries, highlighting convergent, divergent, and synthetic single- and double-tip relay zones

Where overlapping fault tips propagate and connect, the relay zone becomes fully breached, forming a double-tip breached relay fault. This produces a lenticular relay zone, where displacement may be uniform or uneven, depending on local stress and fault maturity. Double-tip breached relay faults are identified, indicating advanced fault linkage and strain localisation (see Fig. 7.9).

The distinction between "unbreached" and "overlapping" faults is based on their geometric expression and displacement behaviour. Unbreached faults accommodate displacement along a single, continuous fault plane. They do not interact with neighbouring fault segments, and their displacement patterns remain internally consistent along strike. Unbreached faults record relatively simple kinematic behaviour and preserve the primary deformation signal of the extensional system. In contrast, overlapping faults involve two or more fault segments that interact without immediate through-going linkage.

Displacement is redistributed among these segments, producing complex, spatially variable displacement patterns across the overlap zone. These interactions commonly generate relay zones, curved fault traces, and displacement gradients that reflect progressive fault growth and stress redistribution (see Fig. 7.10).

The conceptual model of the relay fault system, from unbreached overlap through single-tip breaching to double-tip linkage, is seen in Fig. 7.11. This progression shows how relay zones function as transient structures that regulate displacement transfer before evolving into continuous, through-going faults.

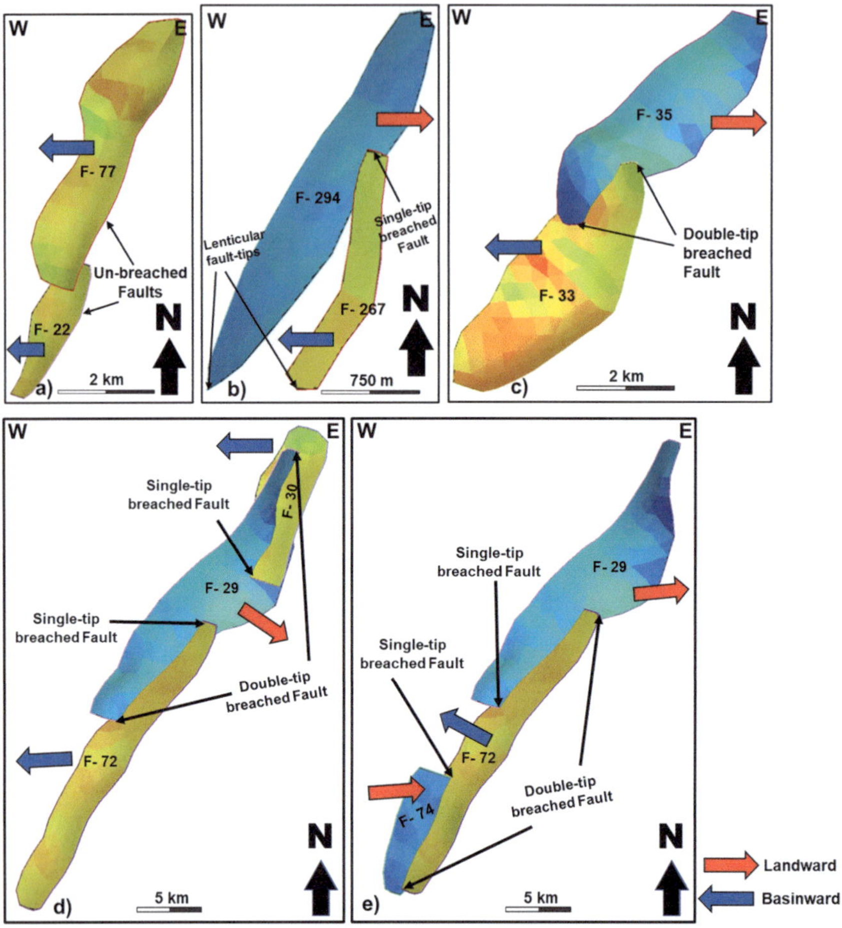

Fig. 7.10 3D perspective view illustrating various types of breached relay faults. **a** Unbreached relay fault (F-22 and F-77). **b** Single-tip breached relay fault (F-294 and F-267). **c** Double-tip breached relay fault (F-33 and F-35). **d** Mixed sequences of single- and double-tip breached relay faults showing Synthetic-Antithetic-Synthetic patterns (F-72, F-29 and F-30). **e** Mixed sequences of single- and double-tip breached relay faults showing Antithetic-Synthetic-Antithetic patterns (F-74, F-72, and F-29). The red arrow indicates landward faults, whereas the blue arrow indicates basinward faults

Understanding the geometry and evolution of relay faults is critical for reconstructing fault growth histories, assessing displacement transfer, and predicting structural connectivity within extensional tectonic systems.

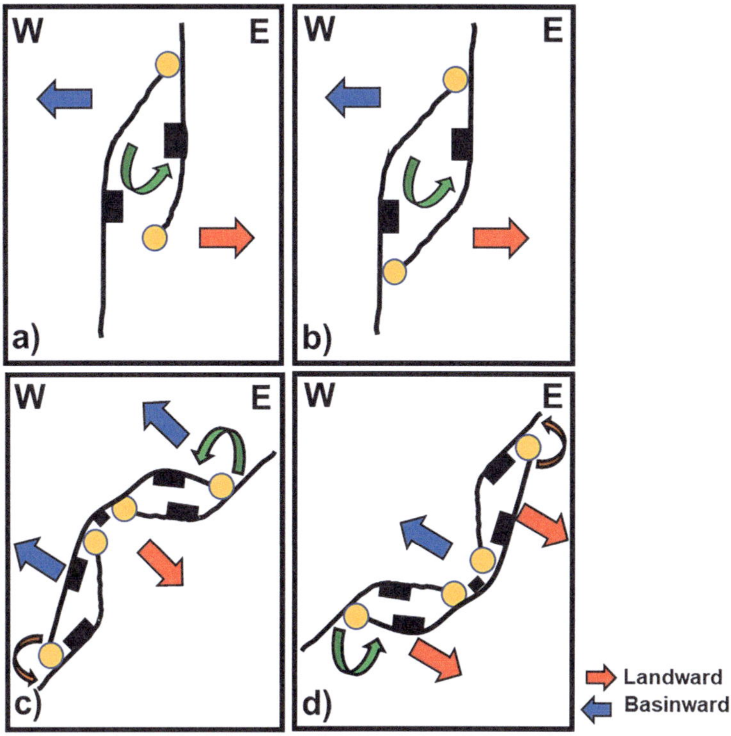

Fig. 7.11 Conceptual model showing various types of breached relay faults. **a** Single-tip breached relay faults; **b** Double-tip breached relay faults; **d** Mixed sequences of single- and double-tip breached relay faults showing Synthetic-Antithetic-Synthetic patterns (relay faults F-72, F-29 and F-30). **e** Mixed sequences of single- and double-tip breached relay faults showing Antithetic-Synthetic-Antithetic patterns (relay faults F-74, F-72 and F-29). The arrows indicate the displacement direction of the faults, with synthetic faults in blue and antithetic faults in red. The green and brown curved arrows show the breached zones

7.3.2 Approaching Faults

Approaching faults are fault segments that propagate towards each other but do not form a direct link. They either stay isolated or interact through stress without intersecting. Two main types of approaching faults are recognised.

1. Isolated approaching faults that propagate independently and show no interaction or connection.
2. Interacting approaching faults, where neighbouring faults influence each other's geometry and displacement patterns as they converge yet remain unlinked to one another or to minor connecting faults.

Approaching faults are classified based on the orientation of their displacement directions.

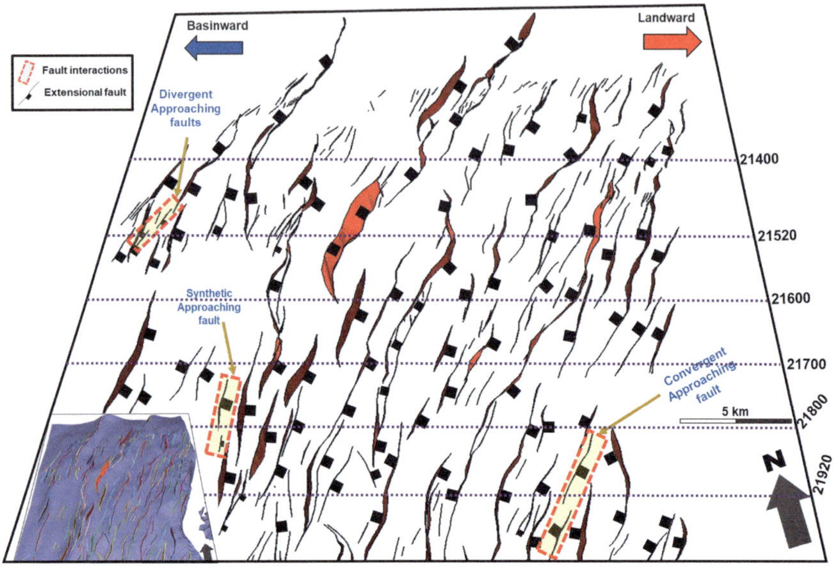

Fig. 7.12 Plan-view fault maps (FM-3) illustrating approaching fault configurations, with fault segments propagating towards one another without direct linkage and recording stress-controlled interaction

- Convergent approaching faults occur where displacements face each other as faults move closer but with opposite senses of displacement (see Figs. 7.12 and 7.13a). This causes stress concentration between fault tips and often reduces displacement in the interaction zone.
- Divergent approaching faults form where displacement directions diverge as they approach, with opposite senses of displacement they approach, with opposite senses of displacement they approach, with opposite senses of displacement (see Figs. 7.12 and 7.13b). Strain disperses from the interaction zone, limiting fault linkage potential.
- Synthetic approaching faults form where displacement directions are the same as the approach direction, with the same sense of displacement they approach, with opposite senses of displacement (see Figs. 7.12 and 7.13c).

Fault map interpretation reveals that convergent, divergent, and synthetic approaching faults are present across the area. They interact with minor subsidiary faults, modifying displacement and tip geometry in response to stress changes, but do not intersect or merge, indicating that propagation stops before linking. Figure 7.14 shows a conceptual model of approaching fault systems with convergent, divergent, and synthetic configurations without linkage. It emphasises that approaching faults are transient interaction geometries that record stress redistribution during fault growth but do not establish full connectivity.

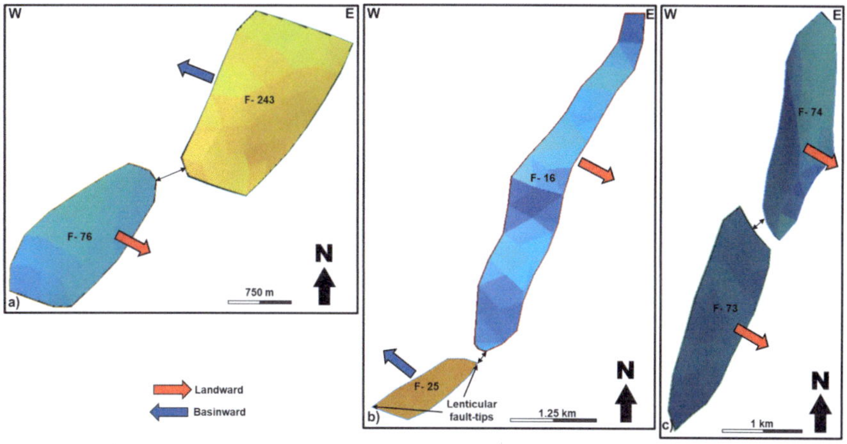

Fig. 7.13 3D perspective view illustrating various types of approaching fault systems. **a** Convergent approaching fault (F-25 and F-16). **b** Divergent approaching fault (F-76 and F-243). **c** Synthetic approaching fault (F-73 and F-74)

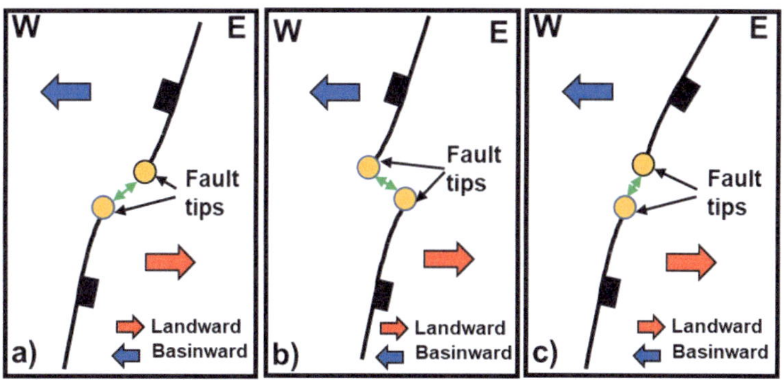

Fig. 7.14 Conceptual model of approaching fault systems. **a** Convergent approaching fault. **b** Divergent approaching fault. **c** Synthetic approaching fault. This system illustrates how faults interact kinematically through stress redistribution while remaining structurally unlinked

7.3.3 Abutting Faults

Abutting faults are fault segments whose tips terminate directly against the fault plane of another structure. These geometries are commonly described as branching faults, reflecting their tendency to form fault networks (Groshong, 2006; Peacock et al., 2017). Abutting faults may develop during the same deformation phase as the main fault or during separate tectonic episodes. They can develop independently and later interact or originate as splays from a pre-existing fault. In all cases,

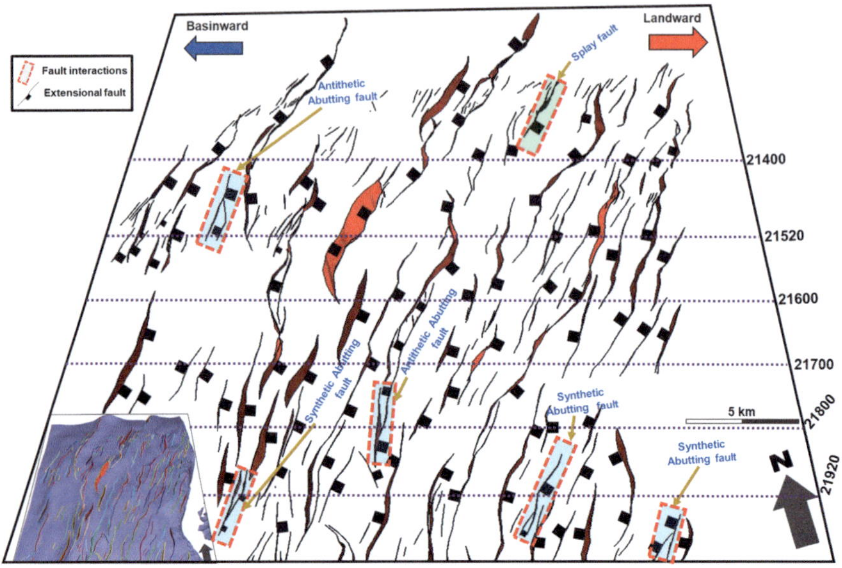

Fig. 7.15 Plan-view fault polygon (FM-3) illustrating abutting and splay fault geometries, highlighted in red broken rectangles, with the inserted Fault Map

displacement transfers along branch lines, indicating mechanical interaction and stress redistribution.

Interpretation of fault maps reveals synthetic and antithetic faults that abut the hanging wall or footwall of a principal fault (see Fig. 7.15). The abutting faults typically intersect the main fault at acute angles of approximately 30°, regardless of whether they about the hanging wall or footwall block.

A conceptual model of abutting fault geometry shows two faults intersecting at approximately 30° (see Fig. 7.16a–d). As deformation progresses, such geometries may evolve into splaying fault patterns (see Fig. 7.16e) or into configurations in which faults approach and ultimately intersect. These variations reflect progressive fault growth and local reorganisation of the stress field.

Within both the hanging-wall and footwall blocks of the principal faults, abutting faults consistently exhibit fault–fault interaction, regardless of whether the abutting fault is synthetic, antithetic, or locally perpendicular to the main fault (see Fig. 7.15). Four end-member abutting configurations are recognised:

1. Synthetic abutting fault terminating in the hanging wall (see Fig. 7.16a).
2. Synthetic abutting fault terminating in the footwall (see Fig. 7.16b).
3. Antithetic abutting fault terminating in the hanging wall (see Fig. 7.16c).
4. Antithetic abutting fault terminating in the footwall (see Fig. 7.16d).

The principal faults define a regionally extensive NE–SW–striking extensional system. As secondary faults propagate towards these main structures from either the

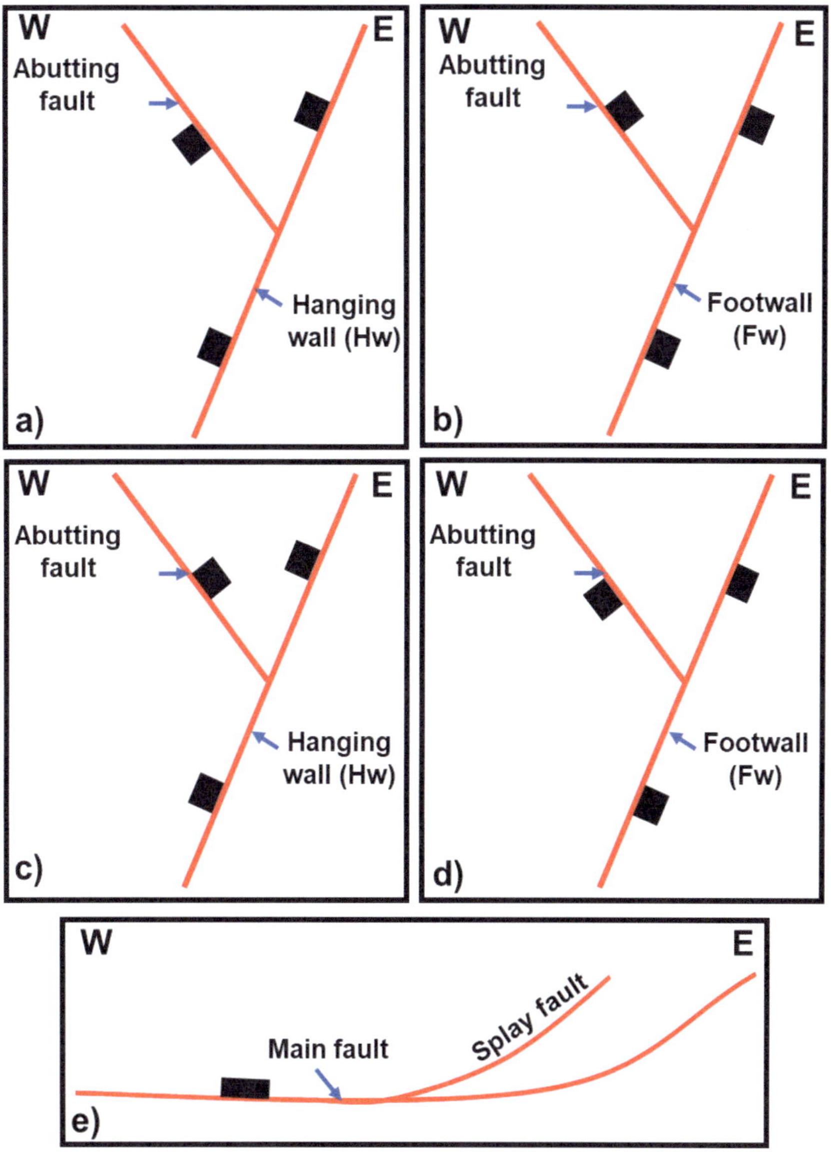

Fig. 7.16 Conceptual models illustrating four abutting and splay fault configurations: **a** synthetic fault abutting the hanging wall, **b** synthetic fault abutting the footwall, **c** antithetic fault abutting the hanging wall, **d** antithetic fault abutting the footwall, and **e** splay fault development, showing bifurcation of a parent fault into subsidiary segments and redistribution of displacement during progressive fault growth

hanging wall or the footwall, they become kinematically linked. During this interaction, the local extension direction rotates to become perpendicular to the approaching fault segment, shifting from NE–SW to NW–SE. Once linkage is established, the system reverts to dominant regional extension.

Abutting faults provide valuable constraints on fault hierarchy, strain partitioning, and stress-field evolution within complex extensional fault networks.

Splay faults commonly develop where a principal fault bifurcates into one or more subsidiary fault segments, forming branching geometries that either link to or diverge from the main fault plane. These splays typically emerge during progressive fault growth and reflect local stress redistribution near the ends of major faults.

Splay faults form near the main fault tips, where displacement decreases along strike and strain divides into multiple subsidiary structures (Needham et al., 1996; see Fig. 7.15). Their expression usually diminishes laterally toward the tip lines, acting as transient pathways for displacement during fault propagation. Splays are bounded by branch lines separating them from the main fault and tip lines where displacement tapers out.

Unlike the parent fault, splay faults can have different orientations from the main fault strike, reflecting local stress variations. This independence lets splays accommodate strain in favourable directions, not necessarily aligned with the regional fault trend.

A conceptual splay-fault model (Fig. 7.16e) helps identify splay faults on maps and in seismic data by showing bifurcation of the fault plane and divergence of segments from a parent fault. Interpretation confirms splay faults (Fig. 7.15), with clear bifurcation and branching. These splays are crucial within the fault network and indicate advanced stages of fault growth.

Splay faults provide key evidence for progressive fault propagation, strain partitioning, and the hierarchical organisation of fault systems within extensional tectonic settings.

7.3.4 Cross-Cutting Faults

Cross-cutting faults are fault relationships in which a younger fault transects and displaces an older fault, or in which two faults intersect such that one clearly offsets the other. These geometries provide some of the most robust criteria for determining relative fault timing within complex fault networks.

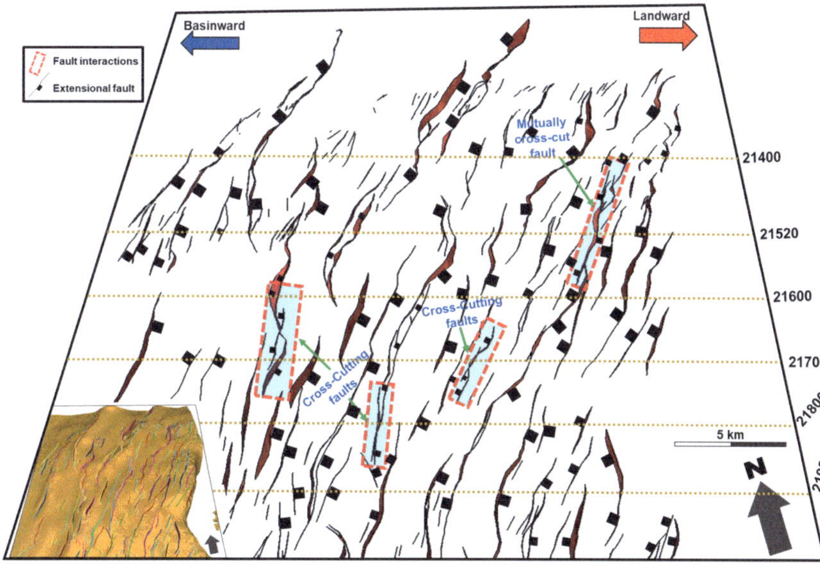

Fig. 7.17 Plan-view fault polygon (FM-4) illustrating cross-cutting fault relationships, highlighted in red broken rectangles, with the inserted Fault Map

Interpretation of the fault maps shows that cross-cutting faults ae present in the Levantine Margin (Fig. 7.17). In the 3D map view, the intersecting fault planes extend vertically along branch lines, clearly showing continuity above and below the intersection (Fig. 7.18). In 2D regional seismic sections, both faults typically extend beyond the branch point, confirming that truncation reflects cross-cutting rather than fault termination.

A conceptualised cross-cutting fault model (Fig. 7.19) shows a fault zone that is later cut and displaced by a younger fault, producing a characteristic trailing segment on the older structure. This geometry records fault reactivation and overprinting during successive deformation events.

Cross-cutting faults, therefore, represent a late-stage interaction mode within the fault network and provide critical constraints on the temporal evolution, kinematic history, and structural overprinting of the extensional system. Table 7.1 shows a comparison of the geometric relationships of fault Interaction types.

Cross-cutting faults are essential markers of fault chronology and play a key role in deciphering the sequence and evolution of deformation in structurally complex extensional terrains.

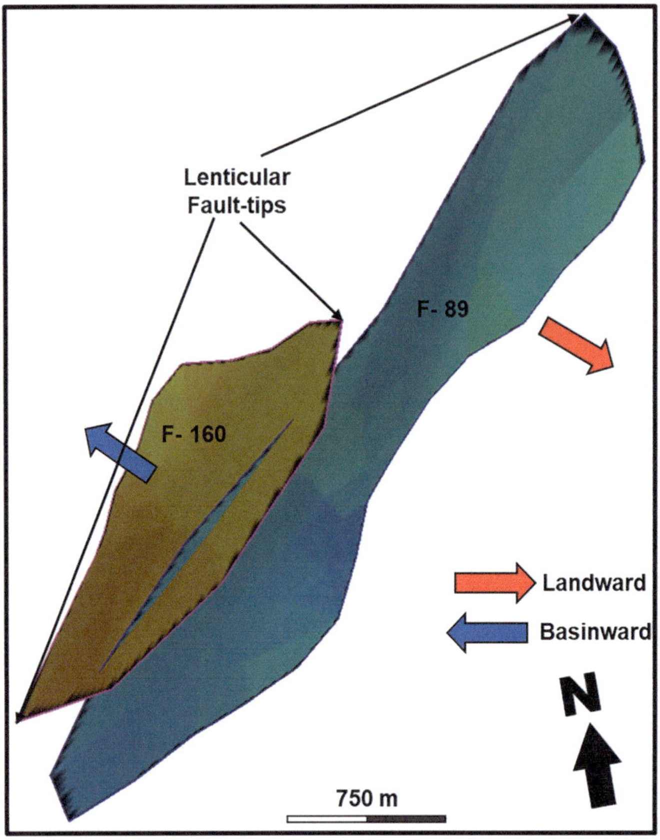

Fig. 7.18 3D perspective view illustrating cross-cutting fault relationships (F-160 and F-89)

7.4 Kinematic Relationships of Interacting Faults

The kinematic relationship between interacting faults is governed by the geometry of the fault–fault intersection line and the directions of displacement on the interacting faults. This relationship provides a direct measure of how slip is transferred, partitioned, or impeded during fault interaction, and is therefore fundamental to understanding fault growth and linkage.

On the Levantine Margin, interacting faults are grouped into three kinematic models, defined by the orientation of the intersection line relative to the displacement direction (Figs. 7.20 and 7.21). These models align with established conceptual models of fault interaction and are recognised across multiple fault maps and seismic interpretations (e.g., Peacock & Sanderson, 1991; Peacock et al., 2017; Walsh et al., 1999; Fig. 7.22).

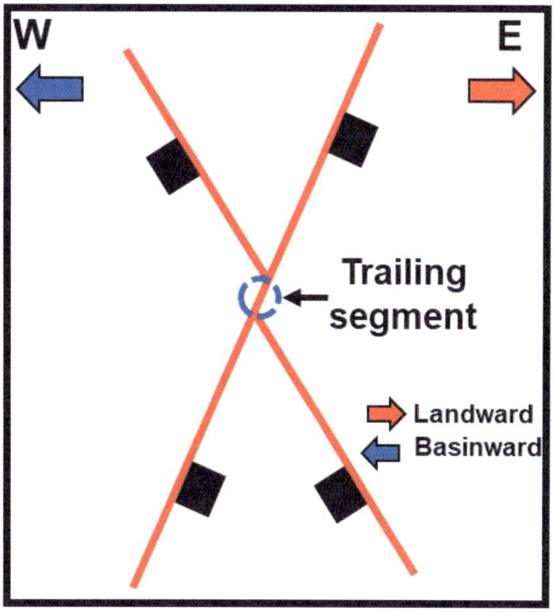

Fig. 7.19 Conceptual model of cross-cutting fault interaction, illustrating displacement of an older fault by a younger fault, formation of trailing segments, and preservation of vertical fault continuity above and below the intersection

Table 7.1 Comparative classification of fault interaction types

Interaction type	Fault relationship	Linkage style	Displacement transfer	Diagnostic geometry	Evolutionary significance
Relay faults	Overlapping, non-intersecting	Indirect (via relay zone)	Progressive, conserved	Overlap with the relay ramp	Early to intermediate linkage stage
Approaching faults	Propagating toward each other	None	Limited, stress-mediated	Convergent/ divergent tips	Arrested growth prior to linkage
Abutting faults	One fault terminates against another	Direct (branch line)	Partitioned	Acute-angle intersection (~30°)	Hierarchical network development
Splay faults	Bifurcation from the parent fault	Direct (branching)	Distributed	Divergent subsidiary segments	Advanced propagation and strain partitioning
Cross-cutting faults	Younger fault offsets older	Overprinting	Disruptive	Trailing segments, truncation	Late-stage fault system reorganisation

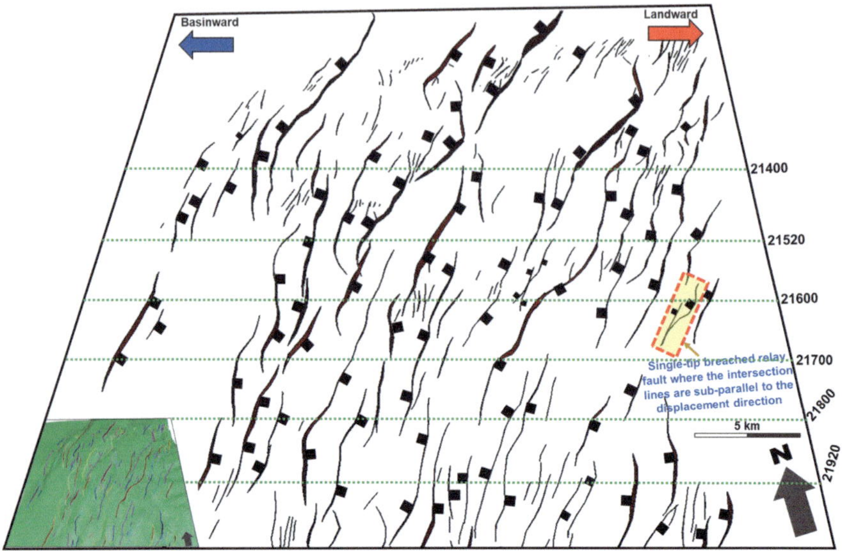

Fig. 7.20 Plan-view fault polygon (FM-2) showing Model 1 kinematic interaction, in which the fault–fault intersection line is parallel to the displacement direction, producing simple, displacement-parallel linkage across a relay ramp, highlighted in red broken rectangles, with the inserted Fault Map

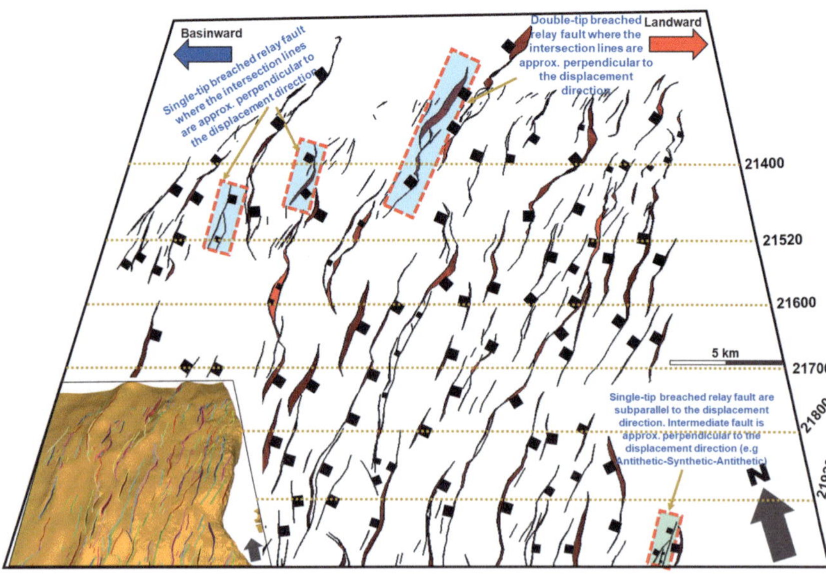

Fig. 7.21 Plan-view fault polygon (FM-4) showing advanced Model 2 and Model 3 kinematic interactions, highlighting increased strain localisation and complex fault connectivity at the Top Messinian level, shown in red broken rectangles with the inserted Fault Map

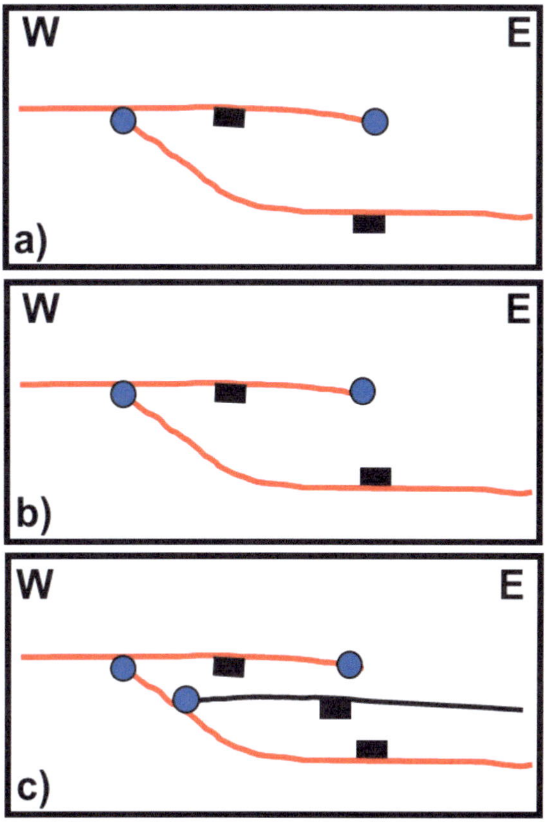

Fig. 7.22 Conceptual models showing the various kinematic relationships between interacting faults. **a** Model 1—kinematic interaction, where displacement vectors are parallel to the fault–fault intersection line, allowing efficient slip transfer with minimal wall-rock deformation; **b** Model 2—kinematic interaction, where displacement is perpendicular to the fault–fault intersection line, requiring significant wall-rock deformation and producing a mechanically complex linkage; and **c** Model 3—mixed kinematic interaction, where a later fault links into a previously breached relay system, producing intersection lines that are parallel to one displacement direction and perpendicular to another

(i) Intersection line parallel to the displacement direction (Figs. 7.20 and 7.22a).
(ii) Intersection line perpendicular to the displacement direction (Figs. 7.21and 7.22b).
(iii) Mixed kinematic relationship (parallel–perpendicular configuration) (Figs. 7.21 and 7.22c).

Model 1: Intersection Line Parallel to the Displacement Direction

In Model 1, the fault–fault intersection line runs parallel to the displacement direction, indicating that two normal faults become kinematically linked by a breaching fault cutting across a relay ramp. This configuration matches the geometry (Fig. 7.20).

In this model, the intersection lines of the breaching faults dip steeply and remain sub-parallel to the displacement directions of the interacting faults. This arrangement produces an Antithetic–Antithetic interaction pattern, in which the linked faults accommodate displacement in opposing dip directions while preserving a coherent transfer geometry across the relay zone (see Figs. 7.20 and 7.22a).

Because displacement aligns with the intersection line, the linkage avoids significant disruption of the adjacent wall rock. As a result, this configuration represents a kinematically simple interaction, characterised by efficient slip transfer and minimal wall-rock deformation (Nelson, 2013; Peacock et al., 2017). This displacement-parallel kinematic relationship is identified in the Levantine Margin.

Model 2: Intersection Line Perpendicular to the Displacement Direction

Model 2 describes a kinematic configuration in which two normal faults interact across a relay ramp, with the fault–fault intersection line oriented perpendicular to the displacement direction. In this setting, fault linkage occurs via single-tip and double-tip breaching faults, producing a more mechanically complex interaction than in displacement-parallel configurations.

For single-tip breaching, the conceptual model shown in Fig. 7.22b closely matches the geometry of the Levantine Margin (see Fig. 7.21). In these cases, one fault tip propagates across the relay ramp and links with the neighbouring fault, while the opposing tip remains arrested. In double-tip breaching, both overlapping fault tips propagate and connect, fully breaching the relay ramp (see Fig. 7.21).

Along the steeply dipping intersecting lines in Model 2, breaching faults link in either Antithetic–Synthetic or Synthetic–Antithetic patterns. Although the primary intersection line is perpendicular to the displacement direction, individual breaching segments commonly develop locally parallel intersection lines, reflecting progressive adjustment during linkage (see Fig. 7.22b).

Model 3: Mixed Kinematic Relationship Across a Relay Ramp

Model 3 describes a mixed kinematic interaction where two faults link across a relay ramp through a single-tip breaching fault, whose trace is nearly parallel to the displacement directions of the breached faults. A third fault then connects to this system. Unlike the breaching fault, this connector fault is roughly perpendicular to the displacement direction. This geometry creates an intersection line parallel to one fault's displacement but perpendicular to the other's (see Figs. 7.21 and 7.22c).

This configuration matches the geometry in Levantine Margin (see Fig. 7.21). The initial linkage occurs along steeply dipping intersection lines and begins as an Antithetic–Antithetic arrangement, with displacement directions on the interacting faults remaining broadly sub-parallel. Subsequent deformation introduces a later fault that intersects and connects the previously breached faults, producing a distinctive composite pattern. The intersection line of this latter connector fault is synthetic to one fault while remaining perpendicular to the displacement directions of the breached system, thereby generating a mixed-slip transfer geometry.

Although the conceptual model depicts crisp geometries, the intersection lines may become curved (see Fig. 7.22c), reflecting progressive growth, local stress rotation, and uneven displacement transfer as the fault network reorganises. Table 7.2 compares the kinematic relationships among fault interaction types.

Table 7.2 Comparative kinematic models of fault interaction

Model	Intersection line versus displacement	Linkage style	Mechanical efficiency	Dominant Fault pattern	Stratigraphic level
1	Parallel	Relay ramp breaching	High	Antithetic–Antithetic	FM-2
2	Perpendicular	Single- & double-tip breaching	Moderate–low	Antithetic–Synthetic	FM-2 to FM-4
3	Mixed	Multi-stage linkage	Low	Antithetic–Synthetic–Antithetic	FM-4 (Top Messinian)

Model 1 recognises a simple, displacement-parallel linkage in map view. Model 2 offers a diagnostic framework for detecting complex relay ramp breaching under unfavourable kinematic conditions. Model 3 describes the transition from simple relay breaching to a multi-stage, mixed-kinematic linkage, where later faulting reorganises earlier relay systems into a connected fault network.

7.5 Displacement and Strain of Interacting Faults

The displacement and strain behaviour of interacting faults is controlled by their relative shear sense, which governs how slip is transferred and how strain accumulates in the surrounding wall rocks. In the Levantine Margin, interacting faults consistently exhibit two fundamental kinematic relationships based on displacement sense: antithetic and synthetic.

These relationships provide a direct framework for interpreting strain accommodation, space creation or loss, and fault linkage efficiency within the fault network.

7.5.1 Synthetic Relationship

Synthetic relationships occur where interacting faults share the same sense of displacement and are generally parallel to the intersection line. In this arrangement, slip vectors are kinematically compatible, enabling more efficient transfer of displacement between faults.

Interpretation of the 3D fault maps indicates that faults in synthetic relationships are displaced approximately parallel to the intersection line, while maintaining consistent displacement senses relative to one another. Synthetic fault interactions

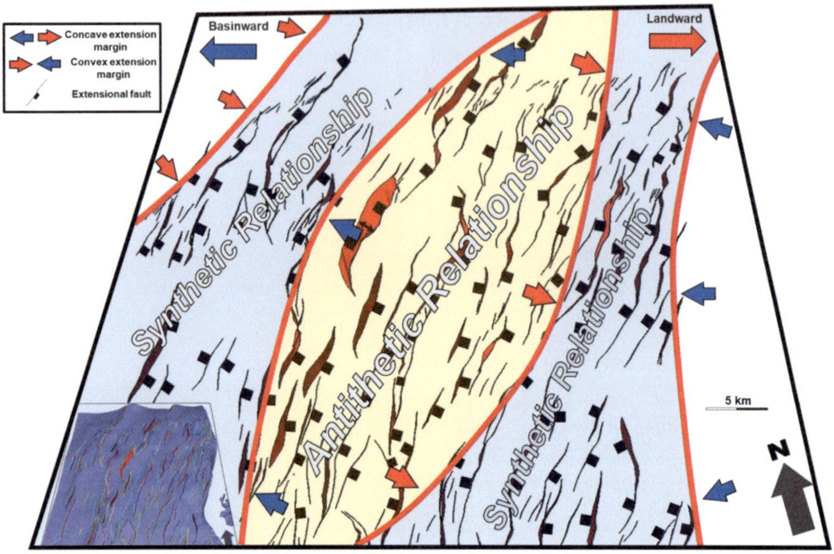

Fig. 7.23 Plan-view fault polygon (FM-3) illustrating displacement and strain relationships between interacting faults, highlighting the regional compatibility of antithetic and synthetic relationships of dominant acute bisectors. Fault arrays exhibit convex-shaped synthetic relationships and concave-shaped antithetic relationships. These fault arrays range from the synthetic relationship, which appears sparse, irregular, and thin, to the antithetic relationship, which appears to expand and exhibit a more regular pattern. The red arrow denotes movement towards the landward direction, while the blue arrow indicates movement towards the basinward direction

are well developed and clearly recognised in the Levantine Margin. The synthetic relationship displays a convex-thinning margin, with the dominant faults exhibiting the same displacement directions (Fig. 7.23).

Because displacement is compatible across the interacting faults, synthetic relationships typically involve lower strain in the wall rocks and promote progressive fault linkage with minimal disruption to surrounding strata. The conceptual model of synthetic fault interaction is shown in Fig. 7.24.

7.5.2 Antithetic Relationship

Antithetic relationships develop where interacting faults exhibit opposite senses of displacement relative to one another and are typically oriented perpendicular to the fault–fault intersection line. In this configuration, slip vectors on the interacting faults diverge, producing a mechanically incompatible geometry.

Analysis of the 3D fault maps shows that faults in antithetic relationships, represented by thick blue spheres, exhibit opposing displacement senses and are displaced approximately perpendicular to the intersection line. This relationship indicate that

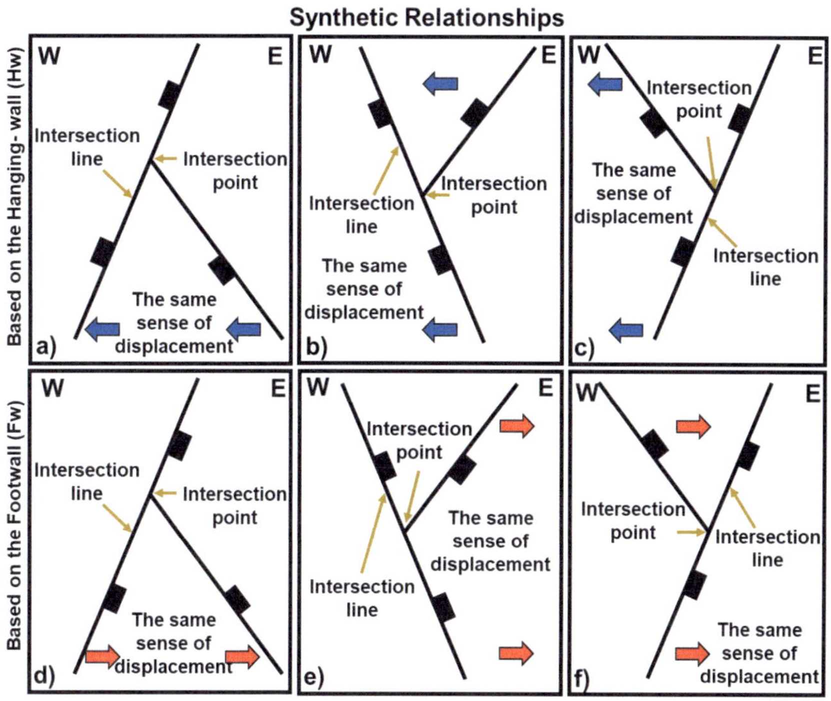

Fig. 7.24 Conceptual model of synthetic fault interaction, showing displacement vectors, efficient slip transfer, and reduced strain accumulation on the **a–c** hanging wall and **d–f** footwall

antithetic interaction is a widespread feature of the fault system (Fig. 7.23). The antithetic relationship shows a concave margin, with the dominant faults exhibiting opposing displacement directions.

Displacement on antithetic faults is kinematically misaligned, causing spatial incompatibilities like local overlaps or voids. These are accommodated by wall-rock deformation, resulting in increased strain, fault-tip bending, and local volumetric strain (Ferrill et al., 2009). The model of antithetic fault interaction is shown in Fig. 7.25. Table 7.3 compares displacement and strain behaviour.

Understanding synthetic and antithetic displacement relationships is essential for evaluating strain localisation, fault connectivity, and the mechanical evolution of fault networks in extensional tectonic settings.

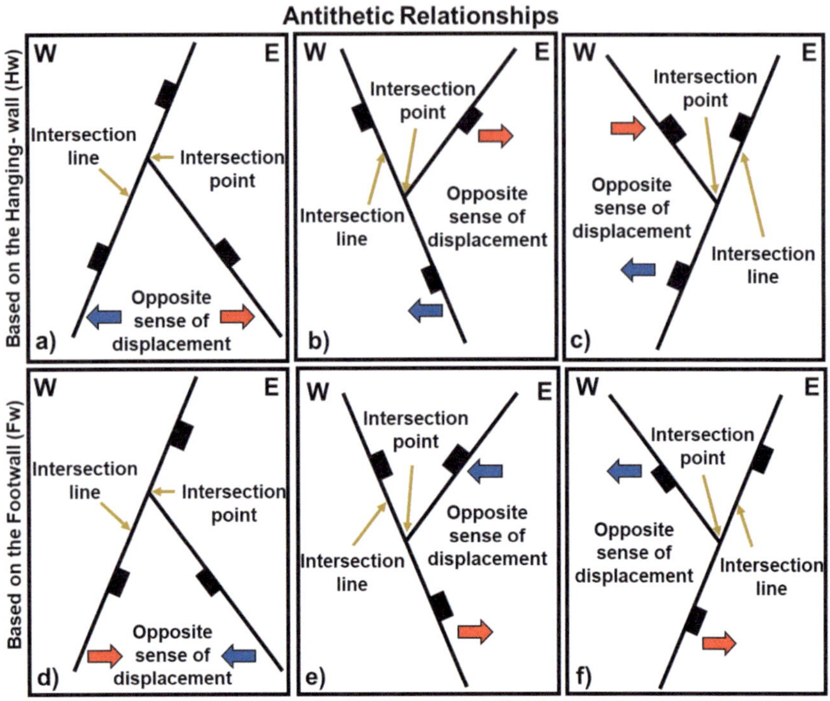

Fig. 7.25 Conceptual model of antithetic fault interaction, illustrating opposing displacement senses, based on the displacement and strains between the interacting faults on the **a–c** hanging wall and **d–f** footwall

Table 7.3 Comparative displacement relationships

Relationship type	Displacement sense	Kinematic compatibility	Strain requirement	Structural outcome
Synthetic	Same sense	High	Low	Efficient linkage
Antithetic	Opposite sense	Low	High	Localised deformation

7.6 Conclusion

This chapter shows that extensional faults on the Levantine margin evolve primarily as interacting, segmented networks rather than as isolated structures. 3D seismic interpretation shows that fault growth is inherently progressive and interaction-driven, controlled by the combined influence of geometry, kinematics, and strain accommodation across multiple stratigraphic levels.

Map-view analyses show that displacement transfer is most efficiently achieved through relay zones, which act as transient structures that regulate fault linkage. The progression from unbreached relays to single- and double-tip breached configurations records increasing strain localisation and mechanical coupling between neighbouring fault segments. Beyond relay geometries, approaching, abutting, splay, and cross-cutting faults further illustrate how stress perturbations around fault tips promote hierarchical fault organisation and network complexity.

Kinematic evaluation of the orientation of fault–fault intersection lines relative to the displacement direction provides a framework for interpreting linkage efficiency and mechanical compatibility. Simple displacement-parallel linkages enable efficient slip transfer, whereas displacement-perpendicular and mixed kinematic relationships require wall-rock deformation and record more complex stages of fault system evolution. Synthetic and antithetic displacement relationships control strain distribution, creating, losing, or redistributing space within interaction zones.

It confirms that extensional fault systems develop as coeval, 3D linked networks under a regional stress field. Integrating geometric, kinematic, and strain-based analyses enables reconstructing fault growth and predicting structural connectivity in rifted continental margins and other extensional tectonic settings.

Learning Objectives

After completing this chapter, readers will be able to:

1. Explain the mechanical principles governing fault interaction.
2. Identify abutting, overlapping, cross-cutting, and splay fault relationships.
3. Analyse relay ramps and breaching mechanisms in extensional systems.
4. Evaluate how strain is transferred between interacting fault segments.
5. Interpret fault linkage patterns in both map view and cross-section.

References

Bull, J. M., Barnes, P. M., Lamarche, G., Sanderson, D. J., Cowie, P. A., Taylor, S. K., & Dix, J. K. (2006). High-resolution record of displacement accumulation on an active normal fault: Implications for models of slip accumulation during repeated earthquakes. *Journal of Structural Geology, 28*(7), 1146–1166.

Ferrill, D. A., Morris, A., & McGinnis, R. N. (2009). Crossing conjugate normal faults in field exposures and seismic data. *American Association of Petroleum Geologists Bulletin, 93*, 1471–1488.

Fossen, H., & Rotevatn, A. (2016). Fault linkage and relay structures in extensional settings: A review. *Earth-Science Reviews, 154*, 14–28. https://doi.org/10.1016/j.earscirev.2015.11.014

Gawthorpe, R. L., Jackson, C. A. L., Young, M. J., Sharp, I. R., Moustafa, A. R., & Leppard, C. W. (2003). Normal fault growth, displacement localisation and the evolution of normal fault populations: The Hammam Faraun fault block, Suez rift Egypt. *Journal of Structural Geology, 25*(6), 883–895.

Groshong, R. H. (2006). *3D Structural Geology: A Practical Guide to Quantitative Surface and Subsurface Map Interpretation*. Springer.

Higgins, S. (2007). 3D *Seismic analysis of the geometry and development of a deep-water fold and thrust belt* (Chapter 2, pp. 1–18.). Cardiff University: United Kingdom

Kelly, P. G., Sanderson, D. J., & Peacock, D. C. P. (1998). Linkage and evolution of conjugate strike-slip fault zones in limestones of Somerset and Northumbria. *Journal of Structural Geology, 20*(11), 1477–1493.

Kim, Y. S., & Sanderson, D. J. (2005). The relationship between displacement and length of faults: A review. *Earth Science Reviews, 68*, 317–334.

Larsen, H. (1988). Relay structures in a Lower Permian basement-involved extension system, East Greenland. *Journal of Structural Geology, 10*, 38.

McClay, K. R., & Ellis, P. G. (1987). Geometries of extensional fault systems developed in model experiments. *Geology, 15*, 341–344.

McClay, K., & Khalil, S. (1998). Extensional hard linkages, eastern Gulf of Suez Egypt. *Geology, 26*, Article 563566.

Morley, C. K., Nelson, R. A., Patton, T. L., & Munn, S. G. (1990). Transfer zones in the East African rift system and their relevance to hydrocarbon exploration in rifts. *American Association of Petroleum Geologists Bulletin, 74*, 1234–1253. https://doi.org/10.1306/0C9B2475-1710-11D7-8645000102C1865D

Muraoka, H., & Kamata, H. (1983). Displacement distribution along minor fault traces. *Journal of Structural Geology, 5*, 483–495.

Needham, D. T., Yielding, G., & Freeman, B. (1996). Analysis of fault geometry and displacement patterns. In E. G. Buchanan & D. A. Nieuwland (Eds.), *Modern Developments in Structural Interpretation, Validation and Modelling, Special Publication* (Vol. 99, pp. 189–199). Geological Society.

Nelson, M. A. (2013). *3D Geometry and Kinematics of Non-colinear Fault Intersections (PhD thesis)*. Cardiff University: U.K.

Nicol, A., Walsh, J. J., Watterson, J., & Bretan, P. G. (1995). Three-dimensional geometry and growth of conjugate normal faults. *Journal of Structural Geology, 17*, 842–862.

Nicol, A., Walsh, J., Berryman, K., & Nodder, S. (2005). Growth of a normal fault by the accumulation of slip over millions of years. *Journal of Structural Geology, 27*, 327–342.

Nixon, C. W., Sanderson, D. J., Dee, S., Bull, J. M., Humphreys, R., & Swanson, M. (2014). Fault interactions and reactivation within a normal fault network at Milne Point Alaska. *American Association of Petroleum Geologists (AAPG) Bulletin, 98*, 2081–2107.

Odonne, F., & Massonnat, G. (1991). Volume loss and deformation around conjugate fractures; comparison between a natural example and analogue experiments. *Journal of Structural Geology, 14*(8–9), 963–972.

Peacock, D. C. (2002). Propagation, interaction, and linkage in normal fault systems. *Earth-Science Reviews, 58*, 121–142.

Peacock, D. C., & Sanderson, D. J. (1991). Displacements, segment linkage and relay ramps in normal fault zones. *Journal of Structural Geology, 13*, 721–733.

Peacock, D. C., & Sanderson, D. J. (1994). Geometry and development of relay ramps in normal fault systems. *American Association of Petroleum Geologists Bulletin, 78*, 147–165.

Peacock, D. C. P., & Sanderson, D. J. (1995). Strike-slip relay ramps. *Journal of Structural Geology, 17*, 1351–1360.

Peacock, D. C. P., Nixon, C. W., Rotevatn, A., Sanderson, D. J., & Zuluaga, L. F. (2017). Interacting faults. *Journal of Structural Geology, 97*, 1–22.

Walsh, J. J., & Watterson, J. (1989). Displacement gradients on fault surfaces. *Journal of Structural Geology, 11*(3), 306–6816.

Walsh, J. J., & Watterson, J. (1991). Geometric and kinematic coherence and scale effects in normal fault systems. *Geological Society, London, Special Publications, 56*(1), 193–203.

Walsh, J. J., Watterson, J., Bailey, W. R., & Childs, C. (1999). Fault relays, bends and branch lines. *Journal of Structural Geology, 21*, 1019–1026.

Walsh, J. J., Bailey, W. R., Childs, C., Nicol, A., & Bonson, C. G. (2003). Formation of segmented normal faults: A 3D perspective. *Journal of Structural Geology, 25*, 1251–1262.

Chapter 8
Quantitative Analysis of Fault Interpretation

Abstract This chapter illustrates how quantitative fault interpretation transforms seismic mapping into a predictive analytical framework. Using 3D seismic data of the Levantine Margin, it extracts and analyses fault length, Max. Displacement, Max. Throw, and displacement gradients across four fault maps (FM-1 to FM-4). Power-law scaling relationships, regression analysis in log–log space, coefficients of determination (R^2), and 95% confidence intervals are employed to evaluate fault growth behaviour, uncertainty, and geological reliability. The results reveal differences in fault maturity and structural complexity. FM-1 and FM-2 exhibit strong, well-constrained scaling relationships, reflecting relatively homogeneous fault populations and consistent stress regimes. FM-3 shows evidence of mechanical maturity and displacement saturation, while FM-4 displays considerable scatter linked to reactivation, heterogeneous stress fields, and structural barriers. Displacement–distance profiles further illustrate the evolutionary progression from immature segmented faults to fully linked, mature systems, and finally to complex structures. By integrating statistical rigour with geological interpretation, the chapter demonstrates that fault-scaling metrics can be used diagnostically rather than descriptively. The chapter provides a transferable quantitative framework for tectonic analysis, reservoir characterisation, and subsurface risk assessment in exploration, production, and CO_2 storage.

Keywords Fault scaling · Power-law relationships · Fault growth · Displacement–length analysis · Max. displacement · Regression analysis

8.1 Introduction

Seismic interpretation is not only about mapping faults; it is about quantifying them to provide predictive insight into the subsurface. Quantitative seismic interpretation (QI) integrates geophysical and geological datasets to predict reservoir properties with greater reliability (Aki & Richards, 2002). When contextualised within structural styles, QI becomes a powerful tool for addressing subsurface complexities. It enables

K. Chiadikobi, *Advances in Horizon Mapping and Fault Interpretation*,
https://doi.org/10.1007/978-3-032-23917-4_8

geoscientists to predict reservoir properties with greater accuracy, optimise drilling locations in structurally challenging settings, improve hydrocarbon recovery, and evaluate formations for safe and efficient CO_2 storage. Such integration is essential across both the exploration and production phases of energy development.

Fault attributes, including fault length, maximum displacement, maximum throw, and displacement gradient, were extracted from four fault maps (FM-1 to FM-4) interpreted from high-resolution 3D seismic datasets. These fault parameters were then correlated and analysed using established power-law relationships (e.g., Marrett & Allmendinger, 1992; Fossen & Rornes, 1996; Needham et al., 1996; Yielding et al., 1996; Gillespie et al., 1992; Gudmundsson et al., 2010; Dawers et al., 1993).

To strengthen the analysis, additional statistical techniques were applied, including: (i) regression fit analysis to evaluate scaling trends, (ii) coefficient of determination (R^2) interpretation to measure correlation strength, and (iii) 95% confidence intervals (CI) to constrain uncertainty and assess reliability.

Through this combined approach, it not only quantifies fault geometry but also examines its scaling behaviour, offering insights into fault growth and the tectonic evolution of the Levantine Margin.

8.2 Statistical Techniques of Fault Analysis

A range of statistical techniques was employed to evaluate the scaling behaviour of faults and to ensure robust interpretation of structural attributes. These techniques provide a quantitative basis for linking observed fault geometries with predictive models of fault growth and deformation.

8.2.1 Power-Law Relationships in Fault Scaling

The power-law relationship is a widely used statistical model for describing the correlation between fault length (L) and dependent fault parameters, such as maximum throw (Tmax) or maximum displacement (Dmax). The relationship is typically expressed as:

$$Y = a.L^n$$

Or in logarithmic form:

$$\log(Y) = \log(a) + b.\log(L)$$

where:

L fault length,

Y dependent variable (either *Tmax* or *Dmax*),
a scaling constant (intercept in log–log space),
b scaling exponent (slope in log–log space).

This formulation enables geoscientists to quantify the relationship between fault growth (lengthening) and displacement (throw). The logarithmic form is particularly useful because it allows a linear regression in log–log space, facilitating statistical evaluation (e.g., R^2 correlation and confidence intervals).

8.2.2 Regression Analysis in Log–Log Space

To evaluate the scaling relationships, linear regression was performed in log–log space using least-squares fitting. This approach enabled the determination of both the scaling exponents (slope) and the intercepts (constant a) of the regression equations.

8.2.3 Coefficient of Determination (R^2)

The reliability of each regression fit was assessed using the coefficient of determination (R^2), which quantifies goodness of fit. Higher R^2 values indicate that the observed data closely conform to the predicted power-law trend, whereas lower values indicate greater scatter and potential structural complexities, such as fault segmentation or linkage.

8.2.4 Confidence Intervals in Scaling Analysis

To assess the statistical robustness of the regression models, 95% confidence intervals (CI) were computed for the regression lines. These intervals are visualised as shaded envelopes in log–log space, indicating the uncertainty associated with the derived scaling relationships.

- Narrow confidence intervals indicate high confidence in the regression fit, reflecting consistent data behaviour and reliable scaling parameters.
- Broader confidence intervals indicate greater data scatter, signalling greater variability in fault attributes and uncertainty in the scaling exponents and intercepts.

Thus, confidence intervals provide not only a statistical measure of model reliability but also a geological signal, as broader scatter may reflect structural complexities such as segmentation, linkage, or salt-related deformation.

8.2.5 Displacement Gradient

The displacement gradient quantifies the spatial rate of change of displacement (slip) along a fault. It is a fundamental parameter for understanding fault growth, segmentation, and mechanical behaviour within the lithosphere, as it indicates how displacement accumulates and tapers, providing a diagnostic tool for assessing the maturity, linkage, and connectivity of fault segments.

A centralised displacement gradient approach was used, defined as:

Centralised Displacement Gradient $\frac{Dmax}{0.5L}$.

Where:

- Dmax = maximum displacement,
- L = total fault length,
- 0.5L = half-length, assuming symmetry from the fault centre to its tips.

This formulation differs from the conventional expression Dmax/L because dividing by half the fault length effectively doubles the gradient, resulting in a steeper slope (see Fig. 8.1).

The centralised approach offers several advantages:

- It reflects the assumption of symmetric displacement tapering from the fault centre towards both tips.

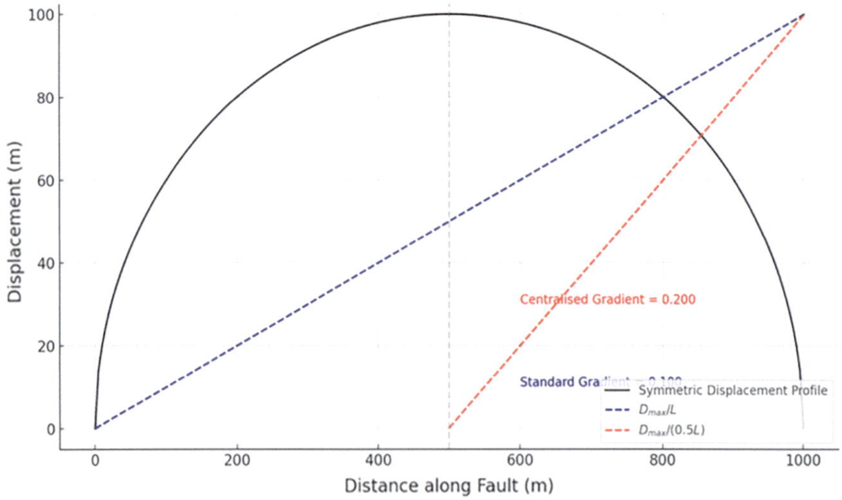

Fig. 8.1 Conceptual diagram illustrating the symmetric displacement profile of a fault (black curve) and the graphical representations of two commonly used gradient estimators: the standard displacement gradient (Dmax/L; blue dashed line) and the centralised displacement gradient (Dmax/0.5L; red dashed line). The latter provides a steeper, more representative measure of fault-strain accommodation

- It is particularly applicable to elastic dislocation models, analytical slip distributions, and geomechanical simulations.
- It better replicates symmetric elliptical displacement profiles commonly observed in both experimental and natural fault systems (see Fig. 8.1).

By adopting this formulation, the analysis can more accurately capture the curvature and amplitude of slip distributions, better represent fault-growth mechanisms, and strengthen structural interpretations.

The power-law model links fault length to displacement and throw, with R^2 values measuring robustness and distinguishing between simple and complex systems. 95% confidence intervals make regression a diagnostic tool. The displacement gradient (Dmax/0.5L) effectively analyses symmetric slip distributions.

8.3 Fault Displacement and Fault Growth Analysis

Analysing fault displacement against distance provides critical insights into fault growth, segmentation, and the mechanical behaviour of fault systems. Displacement profiles along fault lengths capture the evolutionary history of fault development, revealing how faults initiate, propagate, and eventually link. The diagrams presented here (see Fig. 8.2a-d) illustrate different evolutionary stages, from highly segmented networks to more continuous fault systems.

In FM-1, the displacement profile shows multiple sharp peaks with irregular spacing (see Fig. 8.2a). This geometry indicates a highly segmented fault system characteristic of an immature fault network, in which segments grow and accumulate displacement independently.

The sharp peaks correspond to recently linked fault segments, where strain concentrates locally, producing abrupt increases in displacement. The profile's irregularity reflects the effects of fault overlap and relay zone formation, highlighting the complex mechanics of early-stage fault linkage. At this stage, faults are actively evolving, with individual segments occasionally interacting but not yet fully integrated into a single continuous structure.

In FM-2, the displacement profile becomes smoother, with broader peaks (see Fig. 8.2b). This pattern marks a transitional stage in fault development, during which previously isolated fault segments have begun to link mechanically, forming longer, more continuous fault structures.

The broad peaks in the profile indicate that displacement is now shared between neighbouring segments, thereby reducing the sharp discontinuities that characterised FM-1. Although some degree of segmentation remains, partial interaction between

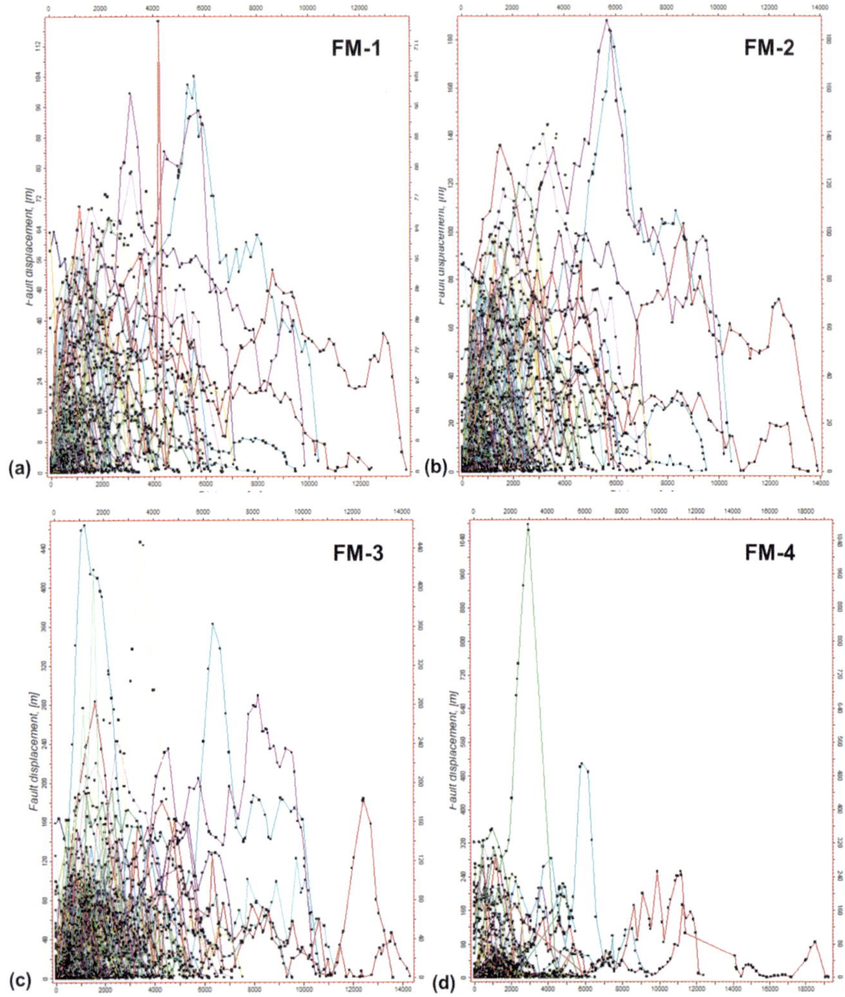

Fig. 8.2 Fault displacement and length analysis. **a–d** Fault displacement profiles along distance for different fault sets, showing variations in displacement magnitudes and their distribution along fault traces of FM-1 to FM-4, respectively. Peaks in displacement correspond to fault segment centres, while displacement decreases towards fault tips

fault tips and overlaps indicates that the system is progressing towards a more mature geometry.

This transitional phase is an important stage in fault growth mechanics, during which linked structures more effectively accommodate strain. As segments coalesce, the fault system becomes increasingly capable of sustaining larger displacements over longer fault lengths.

In FM-3, the displacement profile is characterised by pronounced central peaks forming a bell-shaped curve (see Fig. 8.2c). Maximum displacement occurs near the fault centre and decreases symmetrically towards the tips, a pattern characteristic of mature, fully linked fault systems.

This geometry reflects efficient strain accommodation, with slip distributed more evenly along the fault length. The symmetry of the curve suggests that fault-tip propagation has slowed, indicating that the fault has reached a stage of mechanical maturity. At this point, displacement accumulation is concentrated in the central sections, while the tips accommodate progressively less slip.

Such bell-shaped profiles are widely recognised as hallmarks of well-developed fault systems that have experienced significant slip, marking the culmination of growth through linkage, propagation, and coalescence.

In FM-4, the displacement profile becomes asymmetric, with isolated high-displacement peaks occurring off-centre (see Fig. 8.2d). Unlike the smooth, bell-shaped profiles of mature systems (FM-3), this irregularity indicates a complex fault-growth history shaped by multiple interacting processes.

The off-centre peaks indicate localised strain concentrations, which may arise from: (i) Reactivation of pre-existing faults that disrupt uniform growth, (ii) Heterogeneous stress fields that cause uneven strain distribution, (iii) Structural barriers, such as fault step-overs, bends, or interactions with salt/clay diapirs, that segment and redirect fault growth. This stage reflects a structurally diverse system in which fault segments evolve under varying tectonic and mechanical conditions. As a result, growth rates and displacement patterns vary along strike, producing a fault population that deviates from the classic bell-shaped displacement profile.

The analysis of displacement–distance relationships reveals the dynamic evolution of fault systems. Profiles progress from immature, segmented geometries (FM-1), through transitional linked structures (FM-2), to mature, bell-shaped systems (FM-3), and finally to complex, reactivated patterns (FM-4).

Recognising these evolutionary stages is essential for: (i) Tectonic reconstructions, by revealing how faults initiate, propagate, and link over time; (ii) Geomechanical modelling, by constraining strain distribution, stress interactions, and displacement scaling; and (iii) Applied settings, where fault behaviour informs exploration strategies, hydrocarbon recovery, and the assessment of engineered systems, such as CO_2 storage reservoirs or geothermal sites.

Thus, displacement profiles provide more than geometric information; they are diagnostic tools that record the interplay of fault mechanics, stress regimes, and structural inheritance in both natural and engineered environments. Table 8.1 presents a comparative summary of Fault Growth Stages for FM-1 to FM-4.

Fault growth progresses from immature segmented systems (FM-1) through transitional linkages (FM-2) to mature faults (FM-3). FM-4 captures real-world complexity with irregular displacement indicating reactivation, heterogeneous stresses, and barriers.

Table 8.1 Comparative summary of fault growth stages for FM-1 to FM-4

Fault map	Stage	Displacement profile	Geometric characteristics	Growth stage	Geological significance
FM-1	Immature	Multiple sharp peaks, irregular spacing	Highly segmented; independent fault segments; relay zones	Early growth	Displacement localised; strain partitioned among overlapping segments
FM-2	Transitional	Smoother profiles with broader peaks	Partial linkage; displacement transfer between segments	Linking stage	Strain is accommodated more evenly; segments coalesce into longer faults
FM-3	Mature	Symmetrical bell-shaped curve	Continuous fault; maximum displacement at centre; reduced tip propagation	Mechanical maturity	Efficient strain accommodation; uniform growth; fault system stabilised
FM-4	Complex	Asymmetric profiles with off-centre peaks	Localised strain concentration; irregular displacement	Overprinting stage	Fault reactivation, heterogeneous stresses, and barriers (e.g., salt diapirs) disrupt uniform growth

8.4 Fault Length Versus Maximum Displacement and Maximum Throw

8.4.1 Power-Law Relationship: FM-1

In FM-1, a total of 127 faults traverse the mapped horizon. The average fault length is 2419 m, with notably long faults significantly impacting the horizon. Fault lengths range from 70.15 to 13,782.12 m. The median fault length is 1483.82 m, indicating the presence of shorter faults (see Table 8.2).

The analysis of fault throw shows similar variability. The mean throw is 16.35 m, but outliers significantly inflate this value. The median throw of 9.04 m indicates that most faults are characterised by low throw values (see Fig. 8.3a).

For maximum displacement (Dmax), the average displacement is 226.31 m, driven upward by high values, whereas the median displacement is much lower at 141.60 m. The minimum and maximum values of Max. Displacements are 13.50 m and 1047.70 m, respectively (see Fig. 8.3b).

The scaling relationship shows a positive trend: longer faults accommodate larger displacements (see Fig. 8.2b). However, the scatter of data points reflects statistical

Table 8.2 Summary statistics for fault length, max. throw and displacement for FM-1 to FM-4

Fault map	Parameter	Mean (m)	Median (m)	Std dev (m)	Key observations
FM-1	Fault length	2419	1483.84	2558.80	Dominated by small faults; mature but self-similar growth; few large outliers drive averages
	Max. throw	226.31	141.60	235.10	
	Max. displacement	283.10	186.30	295.70	
FM-2	Fault length	2018	1161.60	2319.53	Greater variability: larger faults are more common but with low throws; transitional strain accommodation
	Max. throw	303.70	143.14	360	
	Max. displacement	375.14	168.30	443.60	
FM-3	Fault length	1831.10	1119.47	2113.40	Highest density, large displacements, and displacement saturation suggest mechanical maturity
	Max. throw	498.80	342.90	487.60	
	Max. displacement	711.32	421.94	858.31	
FM-4	Fault length	2571.10	1581.26	2901.56	Fewer but variable faults; scatter high at short lengths; abundant small faults, rare large ones
	Max. throw	668.45	269.40	816.35	
	Max. displacement	1330	581.40	1892.30	

variability, with clustering at mid-to-large fault values. This distribution suggests self-similar fault growth: many small faults occur, while large faults are less frequent. A few scattered points at low length and low throw represent small-scale faults with minimal slip (see Fig. 8.3b).

Cumulative frequency plots show that frequency decreases smoothly with increasing throw and displacement (see Fig. 7.3c and d). The steep decline at high throw values suggests a relatively high power-law exponent ($D > 2$). Regression fits confirm the robustness of the scaling relationship, with coefficients of determination (R^2) of 0.778 for throw and 0.771 for displacement (see Table 8.2).

In FM-2, a total of 195 faults cut across the mapped horizon. Fault length statistics in FM-2 show a wide range, with an average fault length of 2018 m and a median of 1161.60 m, indicating diversity in fault lengths. The standard deviation is 2319.53 (see Table 8.2).

The analysis of maximum throw shows a significant spread, similar to FM-1. FM-2 exhibits comparable variability to FM-1, with an average maximum throw of about 303.70 m, a median of 143.14 m, and a standard deviation of 360 m (see Table 8.2 and Fig. 8.4a). The maximum displacement in FM-2 also shows substantial variability, with an average of 375.14 m, a median of 168.30 m, and a standard deviation of 443.60 m, indicating a wide range of displacement values (see Table 8.2 and Fig. 8.4b).

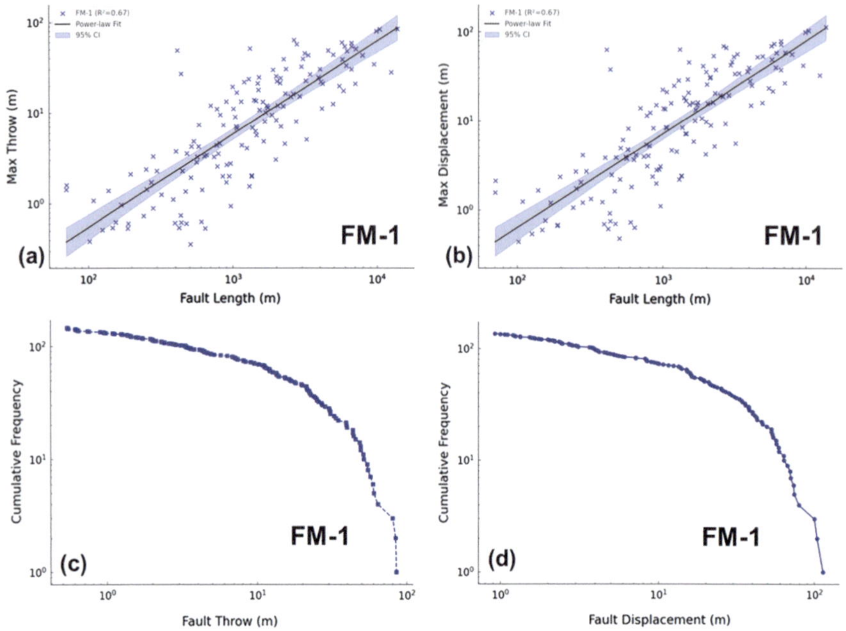

Fig. 8.3 Power-law relationships between **a** fault length and maximum throw and **b** fault length and maximum displacement for FM-1, **c** Cumulative frequency plots of **c** fault throw and **d** fault displacement for FM-1, plotted on log–log scales

These statistics indicate that, as in FM-1, maximum displacement follows a consistent distribution, whereas maximum throw shows a slight deviation, reflecting differences in geological setting and measurement sensitivity. This distinction arises because throw is a subset of displacement and is strongly influenced by fault dip angle: steeper dips accentuate throw, while shallower dips reduce the observed vertical offset.

The scaling profile of FM-2 shows a shallower slope than FM-1, suggesting a higher proportion of larger faults (see Fig. 8.4c, d). However, these larger faults appear to have accumulated less throw, likely due to lower strain accommodation at this stage of development.

In FM-3, 283 faults intersect the mapped horizon, the highest fault density among FM-1, FM-2, and FM-3. Fault lengths vary substantially, with a mean of 1831.10 m and a median of 1119.47 m, indicating a slightly right-skewed distribution. The standard deviation is 2113.40 m (see Table 8.2).

The analysis of maximum throw shows a wide range of values, with an average of approximately 498.80 m and a median of 342.90 m. The standard deviation is 487.60 m (see Table 8.2 and Fig. 8.5a). For maximum displacement (Dmax), FM-3 exhibits the largest value observed to date, with an average displacement of 711.32 m. The median displacement is 421.94 m, and the standard deviation is 858.31 m (see Fig. 8.5b).

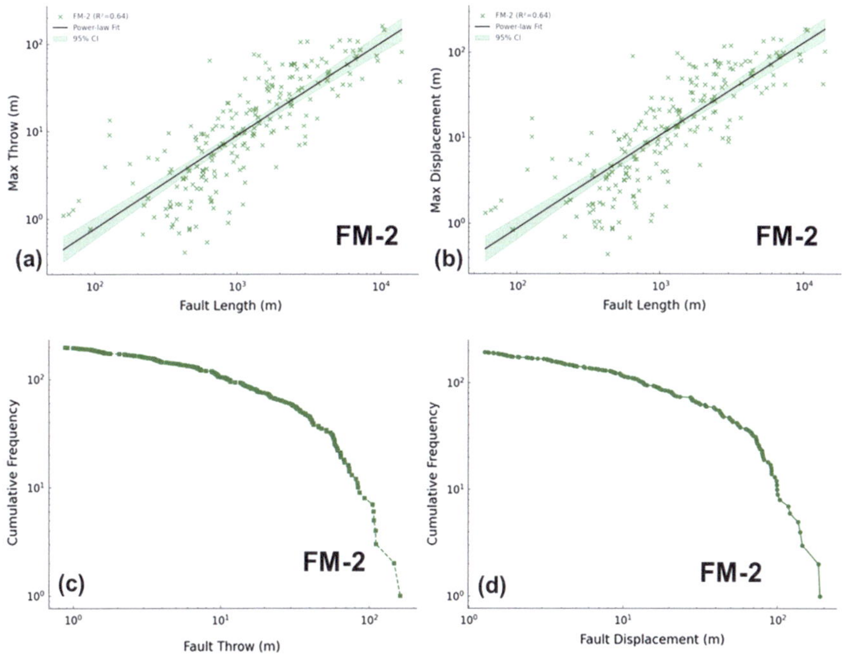

Fig. 8.4 Power-law relationships between **a** fault length and maximum throw, and **b** fault length and maximum displacement, for FM-2. **c** Cumulative frequency plots of fault throw and **d** fault displacement for FM-2, plotted on log–log scales

This dataset confirms that FM-3 hosts faults with considerably higher displacement than FM-1 and FM-2. The scaling relationship between throw and displacement remains broadly consistent across all three fault maps, reflecting the geometric dependence of throw as a subset of total displacement (see Fig. 8.5c, d).

Notably, FM-3 exhibits displacement saturation at longer fault lengths. This suggests a mechanical limit to the accumulation of displacement per unit length, a behaviour often associated with mature fault systems that have approached their maximum strain capacity.

In FM-4, a total of 79 faults cut across the mapped horizon, representing the smallest population among the four fault maps. The fault-length data show a right-skewed distribution, with a mean of 2571.10 m. The median is significantly lower at 1581.26 m, and the standard deviation is 2901.56 m.

The analysis of maximum throw shows a wide range of values, again right-skewed, with an average of 668.45 m. The median maximum throw and standard deviation are 269.40 m and 816.35 m (see Table 8.2 and Fig. 8.6a). For maximum displacement (Dmax), FM-4 exhibits the largest value across all datasets, with an average displacement of 1330 m. The median displacement and standard deviation are 581.40 m and 1892.30 m (see Fig. 8.6b).

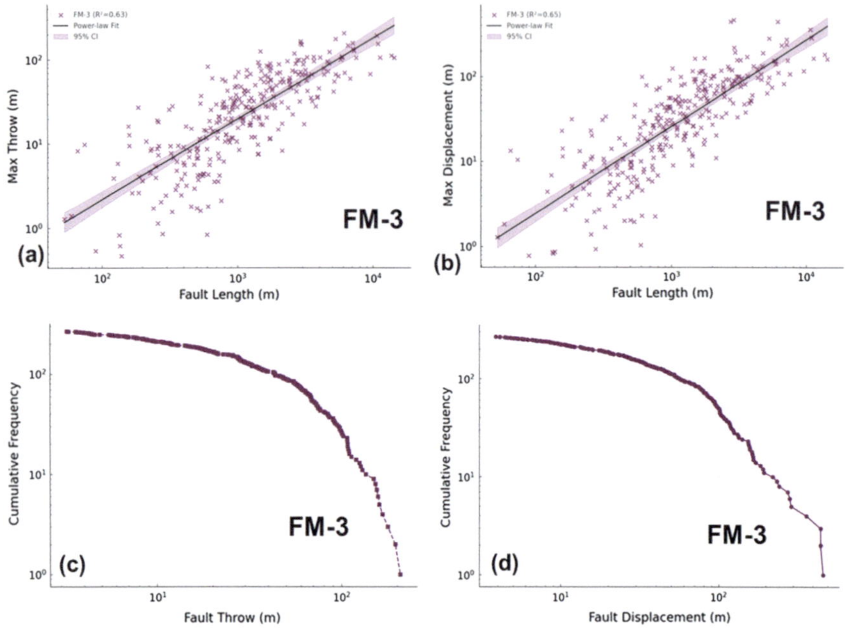

Fig. 8.5 Power-law relationships between **a** fault length and maximum throw and **b** fault length and maximum displacement for FM-3, **c** Cumulative frequency plots of **c** fault throw and **d** fault displacement for FM-3, plotted on log–log scales

The FM-4 dataset is characterised by scattered values, particularly at short fault lengths. The high variance at short lengths suggests heterogeneous growth histories, with some faults in early-stage development and not yet scaling predictably with length. The wide scatter in maximum throw also indicates geological variability, possibly influenced by structural barriers, salt-related deformation, or fault interaction effects (see Fig. 7.6c, d).

Consistent with fracture mechanics principles, FM-4 exhibits a high abundance of small faults and a relative scarcity of large, high-displacement faults. This distribution reflects the natural tendency for faults to initiate as small structures that later grow through propagation and linkage, producing fewer but larger faults.

FM-1 shows a mature, self-similar distribution with small faults. FM-2 shifts to larger faults with lower throws, indicating partial strain. FM-3 is a mechanically mature system with displacement saturation and high fault density. FM-4 exhibits structural complexity, more scatter, many small faults, and rare large, high-displacement faults.

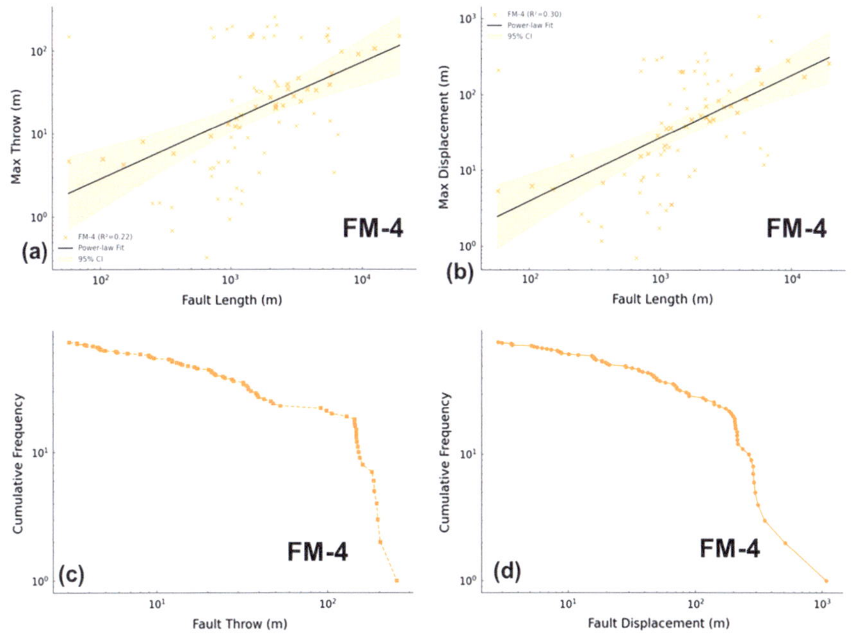

Fig. 8.6 Power-law relationships between **a** fault length and maximum throw, and **b** fault length and maximum displacement, for FM-4. **c** Cumulative frequency plots of fault throw and **d** fault displacement for FM-4, plotted on log–log scales

8.4.2 Regression, Coefficient of Determination (R^2), and Confidence Interval

In FM-1, both scaling relationships, fault length versus throw and fault length versus displacement, show strong statistical consistency. Regression analysis yields R^2 values of 0.62 for both relationships (see Table 8.3 and Fig. 7.3a, b), confirming a robust linear correlation in log–log space. This indicates that fault throw and displacement scale predictably with fault length in this domain.

The regression lines closely follow the overall trend in the data, while the 95% confidence intervals form narrow envelopes around the fits. This narrowness reflects low uncertainty and minimal scatter, suggesting that the dataset is both statistically stable and geologically meaningful. The tight bounds imply that the regression parameters (scaling exponent and intercept) are well-constrained, yielding a model that is both reliable and predictive (see Fig. 7.3a, b).

From a geological perspective, the robustness of the FM-1 power-law fit suggests that this dataset represents a relatively homogeneous fault population. This homogeneity may reflect a single deformational phase or a well-constrained mechanical stratigraphy that governed fault growth with limited overprinting or reactivation.

Table 8.3 Summary of regression performance across fault maps FM-1 to FM-4, including R^2 values for fault length against both maximum throw and displacement, qualitative assessment of confidence interval widths, and interpretive insights into fault population variability

Fault map	R^2 (Max throw)	R^2 (Max displacement)	Confidence interval (CI) width	Scatter behaviour	Geological interpretation
FM-1	0.62	0.62	Narrow, tightly constrained	Minimal scatter: regression closely follows the trend	Homogeneous fault population; consistent growth; single-phase deformation or uniform stratigraphy
FM-2	0.65	0.65	Narrow across the full fault length range	Minimal residual scatter: regression passes through dense data core	Dynamically self-similar faults; consistent stress regime; scaling governed by fracture mechanics principles
FM-3	0.61	0.64	Moderate; broader at large lengths	Pronounced scatter, especially at fault tips	Structural heterogeneity, lithological variations, mixed maturity stages, and displacement saturation limits
FM-4	0.30	0.40	Broad; > 1 order of magnitude at large lengths	High scatter across all sizes; the largest faults deviate strongly	Complex multiphase deformation, reactivation, and variable strain accommodation; salt/ structural barriers disrupt scaling

In FM-2, the scaling relationships between fault length and throw, and between fault length and displacement, are well defined and statistically robust. The regression analysis yields an R^2 of 0.65 (see Table 8.3), slightly higher than that of FM-1, indicating an even stronger correlation in log–log space.

The regression lines intersect the dense core of the data cloud, capturing the central trend with minimal residual scatter. The 95% confidence intervals remain narrow across the entire range of fault lengths, highlighting the reliability of the regression models (see Fig. 7.4a, b). This stability indicates that the scaling behaviour is consistent and not unduly influenced by outliers or irregular fault subsets.

The high degree of coherence across magnitudes suggests that the FM-2 fault population is dynamically self-similar, consistent with expectations from fracture mechanics. Such behaviour implies that the same mechanical processes govern both small and large faults within this domain. Moreover, the uniformity of the scaling relationships indicates that the faults likely developed under relatively consistent stress fields and mechanical conditions, with little evidence of overprinting by multiphase deformation.

In FM-3, the regression models retain a clear power-law trend, though with slightly reduced statistical coherence compared with FM-1 and FM-2. The R^2 values are 0.61

for maximum throw and 0.64 for maximum displacement (see Table 8.3 and Fig. 7.5a, b). Although still robust, these values are marginally lower, indicating increased variability in the dataset.

The data show greater scatter around the regression lines, particularly at larger fault lengths, where the 95% confidence intervals widen. This broadening indicates reduced predictive certainty for the largest faults, many of which deviate markedly from the mean trend. This behaviour highlights the interplay of measurement variability and genuine geological complexity.

The increased uncertainty is likely linked to (i) local structural heterogeneities, including fault linkage and interaction effects; (ii) lithological variations that alter displacement and throw scaling; and (iii) differing fault maturity stages, producing inconsistent growth patterns within the dataset. Thus, while FM-3 retains an overall scaling trend, its regression fits reveal the limitations of simple power-law models when confronted with heterogeneous fault populations.

In FM-4, the regression models show the greatest variability across all datasets. Although a power-law trend remains discernible, the R^2 values are markedly lower than in FM-1 to FM-3: 0.30 for maximum throw and 0.40 for maximum displacement (see Table 8.3 and Fig. 8.6a, b). These values indicate a weak correlation between fault length and kinematic attributes, particularly for displacement.

The 95% confidence intervals are notably wide, with some sections spanning more than an order of magnitude along the x-axis. This level of dispersion indicates substantial uncertainty in the regression parameters and suggests that scaling relationships in FM-4 may be non-linear at larger fault lengths.

The pronounced scatter and variability likely reflect the effects of:

- Structural complexity, including fault linkage, step-overs, and interaction with salt structures.
- Reactivation and multiphase deformation, which overprint simple growth relationships.
- Spatially variable strain accommodation producing inconsistent scaling behaviour.
- Differences in fault connectivity and segment maturity introduce heterogeneity into the dataset.

Despite these complexities, the regression still captures the overall scaling trend, though predictions from FM-4 should be treated with caution. The dataset highlights the limitations of applying idealised power-law models to structurally diverse systems.

The comparative analysis shows that fault scaling relationships are highly dataset-dependent, reflecting variations in fault population characteristics, geological setting, and measurement precision.

- FM-1 and FM-2 provide robust predictive capability, with relatively high R^2 values and narrow confidence intervals, indicating homogeneous fault populations that likely formed under consistent stress and mechanical conditions.

- By contrast, FM-3 and FM-4 show lower regression performance and wider confidence intervals, underscoring the growing influence of structural complexity, fault maturity variations, and multiphase deformation. These datasets emphasise the need to incorporate uncertainty quantification into scaling assessments, particularly in evolutionarily advanced or salt-influenced settings.

Overall, the regression performance across FM-1 to FM-4, summarised by R^2 values, confidence interval widths, and fault population variability (see Table 8.3), indicates that although power-law models are valuable predictive tools, their reliability diminishes in geologically heterogeneous systems.

Fault scaling relationships vary with dataset and geological settings. Early, homogeneous populations (FM-1, FM-2) show strong predictive scaling, but mature, complex systems (FM-3, FM-4) need careful interpretation with uncertainty included in fault growth models.

8.5 Fault Length Versus Centralised Displacement Gradient (FM-1 to FM-4)

The analysis of fault length against the centralised displacement gradient (Dmax/0.5L) across FM-1 to FM-4 shows variable scaling behaviour, characterised by distinct yet weak power-law relationships (see Fig. 8.7a-d). Fault lengths span a wide range (52 m to ~ 19,200 m), while displacement gradients vary from 0.04 to 9.80.

Linear regression in log–log space produced low R^2 values across all datasets (FM-1: 0.022; FM-2: 0.0004; FM-3: 0.0016; FM-4: 0.0053; see Table 8.4), indicating that fault length is only weakly correlated with the centralised displacement gradient.

The regression slopes provide further insight into population behaviour.

FM-1 ($b = -0.11$), FM-3 ($b = -0.03$), and FM-4 ($b = -0.09$) show negative slopes, indicating a mild inverse relationship in which shorter faults exhibit slightly higher displacement gradients.

FM-2 ($b = 0.02$): shows a near-zero slope, suggesting that displacement gradients are scale-independent, with little variation between short and long faults.

These contrasting patterns may reflect differences in structural maturity and strain partitioning within each population. Mature faults (FM-3, FM-4) may accommodate strain more diffusely, resulting in weak gradient scaling, whereas early-stage faults (FM-1, FM-2) preserve more localised variability.

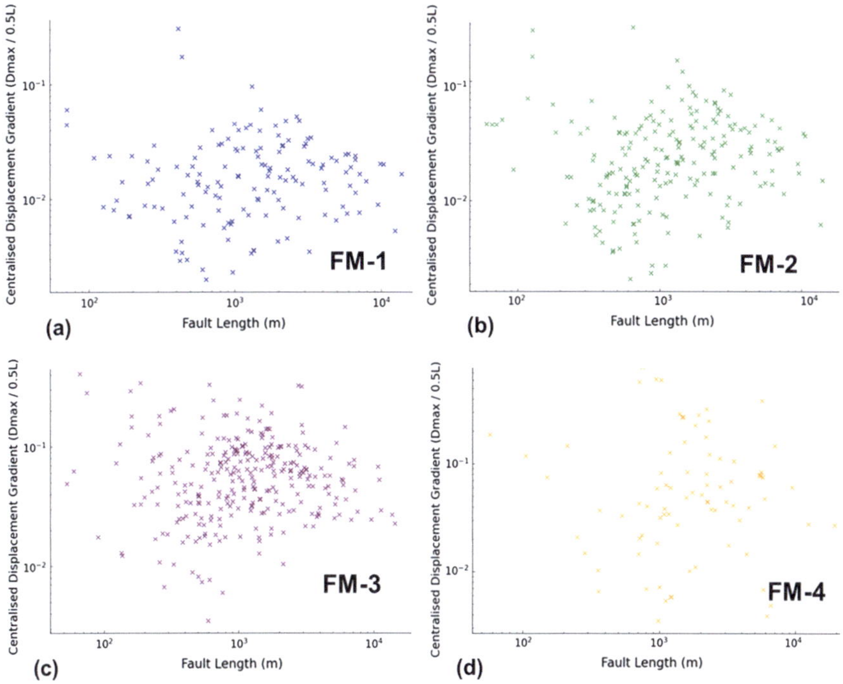

Fig. 8.7 Log–log plots of the centralised displacement gradient (Dmax/0.5L) versus fault length for **a** FM-1, **b** FM-2, **c** FM-3, and **d** FM-4

Displacement gradients from FM-1 to FM-4 show weak correlation with fault length ($R^2 < 0.03$). Shorter faults in FM-1, FM-3, and FM-4 have higher gradients, while FM-2 is more uniform, indicating diverse effects of fault maturity and complexity on displacement gradient scaling.

Table 8.4 Comparative Regression Results for Fault Length vs. Centralised Displacement Gradient (FM-1–FM-4)

fault map	Fault length range (m)	Displacement gradient range (Dmax/0.5L)	Slope (b)	R^2	Interpretation
FM-1	70–13,782	0.05–7.40	–0.11	0.022	Mild inverse trend: shorter faults show higher gradients; weak correlation reflects localised strain accommodation
FM-2	85–15,904	0.04–9.10	+ 0.02	0.0004	Near-zero slope: scale-independent behaviour; gradients broadly consistent across fault lengths, suggesting uniform stress regime
FM-3	60–17,845	0.06–8.60	–0.03	0.0016	Weak negative slope: larger faults distribute displacement more diffusely; this reflects mixed maturity and structural heterogeneity
FM-4	52–19,200	0.07–9.80	–0.09	0.0053	Strong scatter and broad CI: inverse scaling present but unreliable; indicates complex multiphase deformation and variable strain partitioning

8.6 Conclusion

This chapter examined the scaling behaviour of faults on the Levantine Margin through a comprehensive statistical and visual analysis of four fault maps (FM-1 to FM-4). By evaluating relationships among fault length, maximum displacement, maximum throw, and displacement gradient, it provide insights into fault growth, maturity, and mechanical behaviour in an extensional tectonic setting.

The fault length-throw and displacement analyses confirmed that, although power-law trends are broadly evident, their predictive strength varies across datasets. FM-1 and FM-2 produced strong regression fits with relatively high R^2 values (0.62–0.65) and narrow confidence intervals, indicating that fault populations are homogeneous and likely controlled by uniform stress regimes. By contrast, FM-3 and FM-4 showed broader scatter, lower regression performance ($R^2 = 0.30$–0.64), and wider confidence intervals. These results point to increasing geological complexity, including multiphase deformation, fault reactivation, and variable strain accommodation, particularly in salt-influenced or structurally diverse domains.

Analysis of fault displacement profiles revealed the evolutionary stages of fault development. FM-1 displayed irregular, peaked profiles typical of immature, segmented systems. FM-2 showed smoother curves, indicating a transitional stage in which segments begin to link. FM-3 was characterised by symmetric, bell-shaped profiles, indicating that it was mechanically mature and fully linked. FM-4 exhibited asymmetric, irregular displacement distributions, indicating reactivation, stress-field heterogeneity, and structural barriers such as diapirs.

The displacement gradient analysis introduced a centralised formulation (Dmax/0.5L) better suited to symmetric elliptical slip distributions. Across FM-1 to FM-4, correlations between fault length and displacement gradient were consistently low ($r^2 < 0.03$), with mild negative slopes in most datasets. This suggests that shorter faults may sustain higher gradients, whereas larger faults distribute strain more diffusely. FM-2's near scale-independence indicates that uniform stress regimes can override expected gradient-length scaling.

Taken together, the findings show that fault scaling relationships are highly dataset-dependent, governed by statistical variability and geological controls. While robust for early-stage or homogeneous fault populations, scaling laws must be applied with caution in structurally complex settings, where uncertainty quantification is essential. These insights advance our understanding of fault growth mechanisms, reservoir compartmentalisation, and tectonic evolution in extensional margins, with implications for hydrocarbon exploration and CO_2 storage risk assessments.

Learning Objectives:

After completing this chapter, readers will be able to:

1. Measure fault displacement and throw from seismic data.
2. Construct and interpret displacement–length profiles.
3. Analyse displacement gradients to assess fault maturity.
4. Identify deviations from idealised fault growth models.
5. Relate displacement patterns to rift evolution and strain partitioning.
6. Apply statistical methods to analyse fault populations.
7. Evaluate power-law relationships between fault length and displacement.
8. Interpret log–log regression results and confidence intervals.
9. Assess the impact of geological complexity on scaling behaviour.
10. Integrate quantitative scaling results with geological interpretation.

References

Aki, K., & Richards, P. (2002). *Quantitative seismology*. MIT Press.

Dawers, N. H., Anders, M. H., & Scholz, C. H. (1993). Growth of normal faults; displacement-length scaling. *Geology, 21*, 1107–1110.

Fossen, H., & Rornes, A. (1996). Properties of fault populations in the Gullfaks Field, northern North Sea. *Journal of Structural Geology, 18*(23), 179–190.

Gillespie, P. A., Walsh, J. T., & Watterson, J. (1992). Limitations of dimension and displacement data from single faults and the consequences for data analysis and interpretation. *Journal of Structural Geology, 14*(10), 1157–1172.

Gudmundsson, A., Simmenes, T. H., Larsen, B., & Philipp, S. L. (2010). Effects of internal structure and local stresses on fracture propagation, deflection, and arrest in fault zones. *Journal of Structural Geology, 32*(11), 1643–1655.

Marrett, R., & Allmendinger, R. W. (1992). Amount of extension on "small" faults: An example from the Viking Graben. *Geology, 20*(1), 47–50.

Needham, D. T., Yielding, G., & Freeman, B. (1996). Analysis of fault geometry and displacement patterns. *Geological Society, London, Special Publications, 99*(1), 189–199.

Yielding, G., Needham, T., & Jones, H. (1996). Sampling of fault populations using sub-surface data: A review. *Journal of Structural Geology, 18*(2–3), 135–146.

Chapter 9
Comparative Studies of the Levantine Margin with Selected Continental Margins

Abstract This chapter provides a comparative analysis of fault geometries on the Levantine Margin and three analogue continental margins: the offshore western Niger Delta Basin, the Orphan Basin–Flemish Pass, and the North Chukchi Basin–Hanna Trough. Using interpreted 2D seismic profiles and map-view fault datasets, the chapter examines structural styles, fault polarity, and the kinematic relationships of fault interactions to explain how extensional strain is organised and transferred across different tectono-stratigraphic settings. The analysis indicates that detachment depth and rheology primarily control fault geometry, spacing, polarity stability, and interaction style. Evaporite- and shale-decoupled margins display dense fault networks, mixed or domain-controlled polarity, and relay-rich, adaptive kinematics, while crustal- and basement-involved margins show early fault linkage, stable polarity, and vertically coherent deformation. Across all case studies, relay ramps, abutting and splay faults, cross-cutting relationships, and acute bisectors consistently record extension-dominated strain, and their distribution reflects the degree of mechanical decoupling. By integrating geometry, polarity, and kinematics within a unified comparative framework, this chapter moves from descriptive interpretation towards a process-based understanding of rift evolution.

Keywords Fault interaction · Detachment faulting · Kinematic analysis · Shale detachment · Evaporite detachment · Basement-involved rifting · Acute bisectors

9.1 Introduction

This chapter compares fault systems across continental margins, including the Levantine Margin and three global examples: the offshore Western Niger Delta, the Orphan Basin–Flemish Pass, and the North Chukchi Basin–Hanna Trough (see Fig. 9.1). These margins are selected for their diverse fault styles, polarity, and interactions, all of which share an extensional tectonic origin. The chapter uses three interconnected structural descriptors: (i) structural styles (fault geometry and arrangement), (ii) fault polarity (synthetic vs. antithetic), and (iii) fault interactions (relay ramps,

K. Chiadikobi, *Advances in Horizon Mapping and Fault Interpretation*,
https://doi.org/10.1007/978-3-032-23917-4_9

abutments, splay development, cross-cutting relationships). This framework enables comparisons of rift architectures across margins affected by evaporite detachment, shale mobility, and basement extension.

To ensure consistency across case studies, representative 2D seismic sections and map-view interpretations from published studies are reconstructed and re-evaluated within a single interpretive framework. This approach enables direct, scale-consistent comparison of fault spacing, relay geometry, and kinematic compatibility across basins, minimising bias arising from disparate workflows or interpretive philosophies. The comparative analysis draws on established datasets and interpretations from the Niger Delta (Chima et al., 2022), the Orphan Basin–Flemish Pass (Cawood et al., 2021a, 2021b), and the North Chukchi Basin–Hanna Trough (MMS, 2006; Kumar et al., 2011), together with a new structural synthesis of the Levantine Margin (see Fig. 9.1).

A unifying theme is the influence of mechanical stratigraphy on rift architecture. In the Levantine Margin, Messinian evaporites form a ductile shear zone that decouples brittle deformation from deeper crustal extension. Similar decoupling occurs in the Niger Delta over mobile shale and in the Orphan Basin above mid-crustal detachments. By contrast, the North Chukchi Basin–Hanna Trough hosts a rift system characterised by steep, basement-involved faults with limited detachment control. These contrasts help assess how lithology, sedimentation, and extension shape fault development.

Fault interactions are primary indicators of strain evolution, not secondary geometric features. The chapter highlights relay faults, breaching behaviour, abutting relationships, and cross-cutting structures as indicators of displacement and stress changes. Synthetic and antithetic relationships, fault-tip geometries, and acute bisectors indicate extension-driven strain fields. These attributes, when viewed on sections and maps, constrain fault growth history and rift kinematics.

Beyond regional comparison, the chapter provides readers with a transferable interpretive toolkit for frontier basin analysis, hydrocarbon and CO_2 storage assessment, and regional tectonic synthesis. It links geometry, kinematics, and mechanical controls, thereby encouraging a shift from descriptive to process-based structural interpretation.

9.2 Structural Styles

9.2.1 The Levantine Margin

The Levantine Margin features a dense network of normal faults that form a gravity-driven, detachment-controlled continental margin. Faults are evenly spaced, approximately 0.25 to 2 km apart, creating a segmented rift visible on seismic profiles and maps.

Two principal fault geometries characterise the margin:

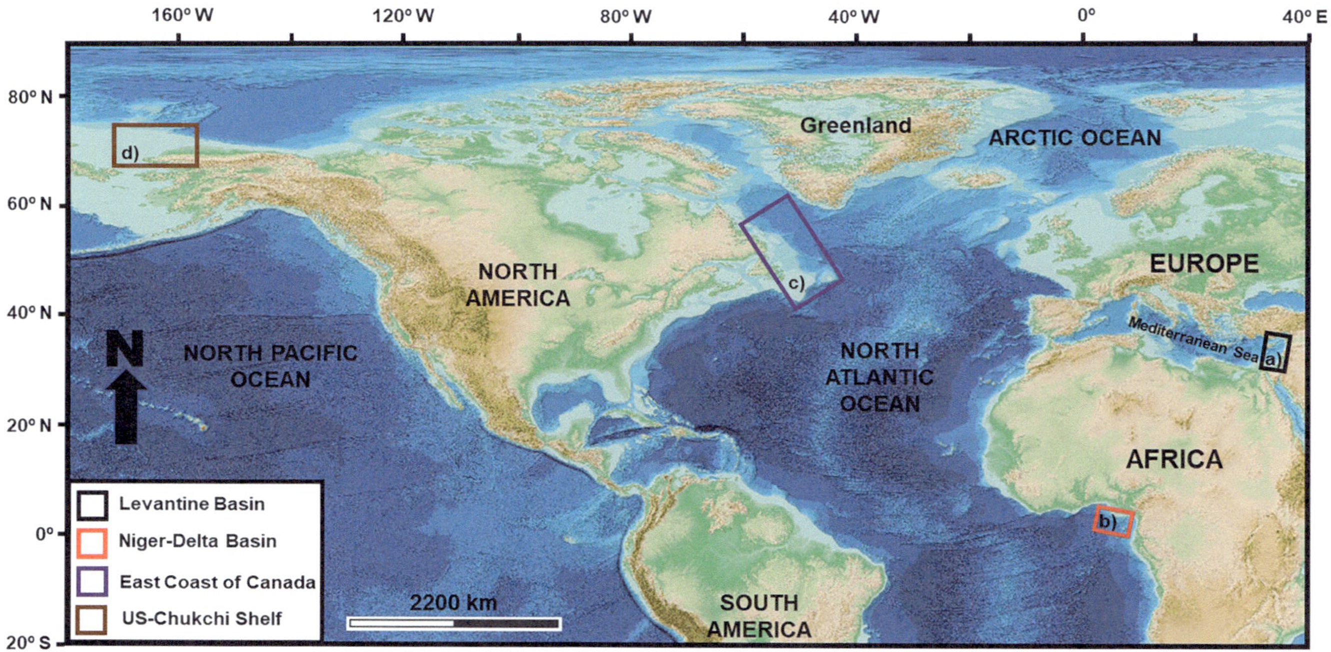

Fig. 9.1 Location map showing **a** the Levantine Margin and other selected areas; **b** the Offshore Niger Delta Basin; **c** the East Coast of Canada, comprising the Orphan Basin and Flemish Pass; and **d** the US-Chukchi Shelf, comprising the North Chukchi Basin and Hanna Trough (adapted from GEBCO, 2022)

- Planar normal faults occur throughout the Plio–Quaternary succession, maintaining near-constant dip angles (see Figs. 6.1a and 6.2).
- Listric normal faults steepen upward and flatten with depth, often rooting in weak horizons within the Messinian succession (see Fig. 6.1b).

These faults form a NE–SW-trending array along the continental slope, constituting a continental margin fault system (Garfunkel and Almagor, 1985; Nelson, 2013). Fault geometry and spacing are governed by gravitational collapse and stratigraphic decoupling rather than by uniform crustal stretching.

Structural style features low-angle detachment faults within Messinian evaporites, forming a ductile shear zone that decouples brittle deformation in the Plio–Quaternary sediments from deeper levels. This efficiently transfers strain between brittle and ductile regimes, enabling high-density faulting in the cover without deep displacement.

Block faulting is common in the area, with tilted fault blocks, half-grabens, and graben geometries. High-bound structures, low-angle planar faults, and listric faults indicate fault rotation during extension (Osmond, 1960; Brun and Choukroune, 1983; Buck, 1991; Park, 1993) (see Fig. 6.4). This architecture is primarily driven by gravitational sliding and sediment loading rather than by uniform lithospheric stretching.

A key diagnostic of the structural style is conjugate fault systems, which appear as V-, X-, and Y-shaped patterns on maps (see Figs. 6.5 and 6.6). These geometries result from interactions between ductile Messinian evaporites, which flow, and the brittle overburden. This produces conjugate fault arrangements that record changes in stress during gravity-driven deformation.

The Levantine margin illustrates weak stratigraphic horizons accommodating extension, with shallow faults above a ductile detachment. This contrasts margins without evaporites and aids in interpreting gravity-driven deformation along salt-bearing margins.

9.2.2 Offshore Western Niger Delta Basin

The extensional domain of the offshore western Niger Delta Basin features gravity-driven structures underlain by a mobile shale detachment and characterised by rapid sediment accumulation. Deformation is primarily confined to a shelf-parallel extensional area, where fault geometry, spacing, and block architecture indicate syn-sedimentary growth rather than uniform lithospheric stretching (Connors et al., 1998; Morley and Guerin, 1996; Graue, 2000; Rouby et al. 2011).

The domain is characterised by widely spaced, basinward-dipping listric normal faults. These faults are steep near the surface and flatten into overpressured shale horizons within the Akata Formation, forming a shale-detachment–controlled extensional structure. Hanging-wall blocks often rotate during fault slip, producing rollover anticlines typical of the Niger Delta's extensional system (Short & Stauble, 1967; Damuth, 1994; Corredor et al., 2005; see Fig. 9.2).

Fault spacing in this extensional domain is greater than at evaporite-controlled margins, such as the Levantine Margin, indicating that extension is accommodated by fewer high-displacement faults rather than dense fault populations. Consequently, deformation is focused into discrete fault-bounded blocks rather than distributed across closely spaced fault arrays.

The structural style shows progressive fault growth over time, with growth strata thickening near active faults and fault geometries indicating incremental rotation of hanging-wall blocks as displacement increases. This produces asymmetric half-graben structures in which accommodation space is concentrated near major growth faults.

In map view, faults within the extensional domain form curvilinear, broadly NE–SW–trending arrays that align with the regional slope and sediment dispersal. Fault traces are segmented, with overlapping tips, indicating lateral growth and propagation (see Fig. 9.3). This segmentation indicates that shale mobility facilitates the progressive propagation and linkage of faults, rather than the early formation of continuous fault surfaces.

Observations from the offshore western Niger Delta Basin show shale mobility mainly drives extensional deformation. High-displacement growth faults develop over sediment loading. Compared to evaporite-controlled margins, this style emphasises detachment rheology's role in rift architecture and fault system organisation.

9.2.3 Orphan Basin–Flemish Pass

The Orphan Basin–Flemish Pass features large, shallowly dipping detachment faults that control basin shape and fault organisation. It exemplifies a detachment-dominated rift, in which extension occurs primarily through low-angle faulting rather than through steep normal faults (see Fig. 9.4).

Most major faults across the seismic sections dip gently, predominantly to the north-west. These detachment faults shape the basin's structure, influencing the locations, geometry, and segmentation of graben and half-graben at multiple scales.

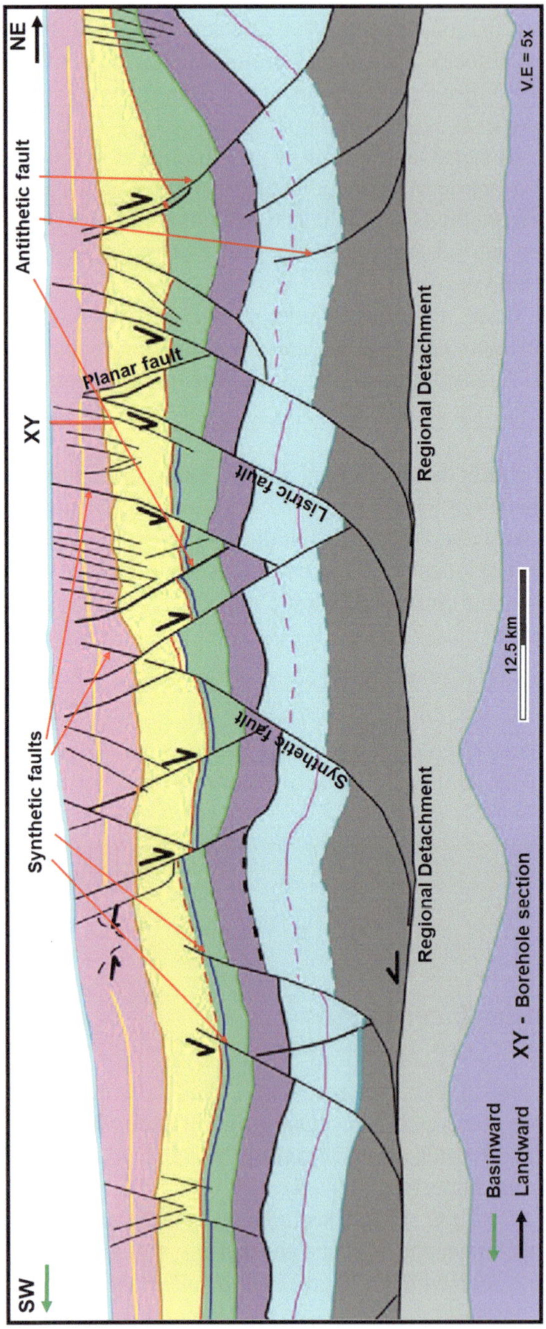

Fig. 9.2 Interpreted regional geoseismic profile of the offshore Western Niger Delta Basin, showing the extensional fault geometry, based on structural styles (listric and planar faults) and fault polarity or displacement directions (synthetic and antithetic faults) (Adapted from Chima et al. 2022)

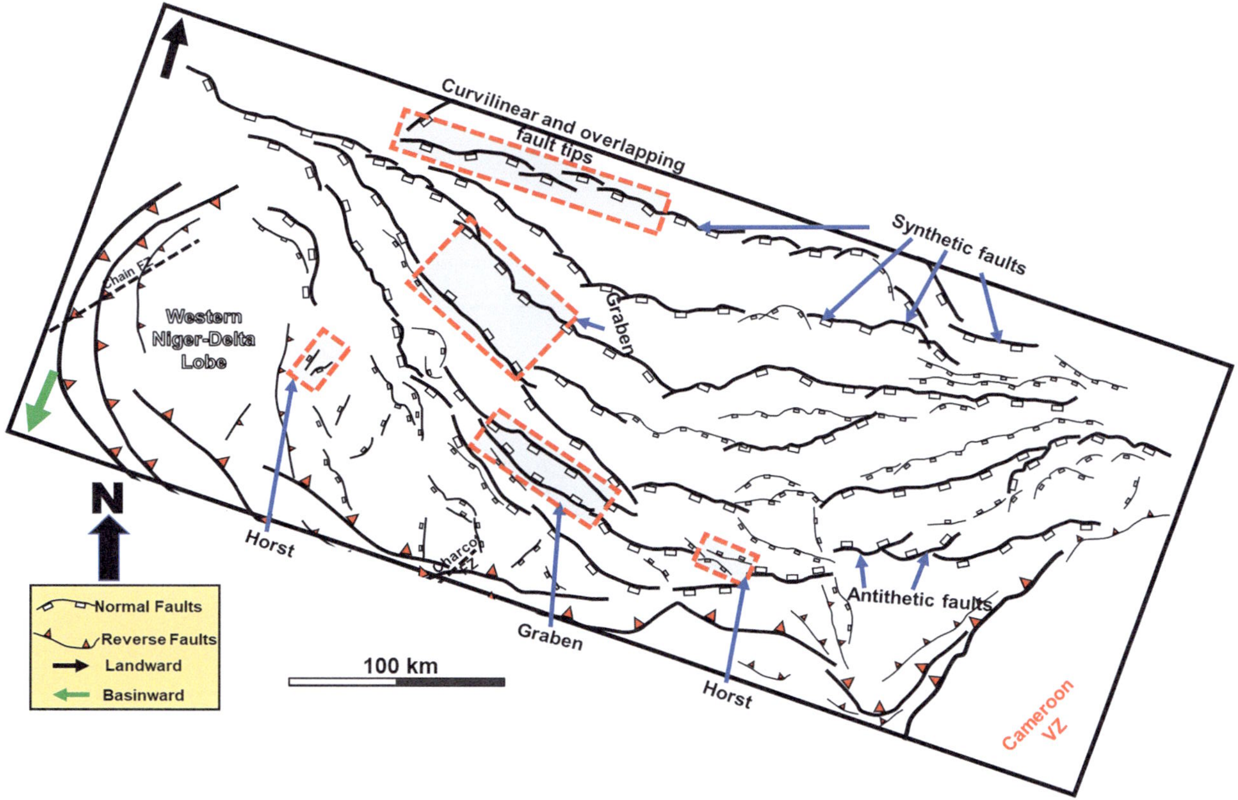

Fig. 9.3 Map view of the offshore Niger Delta Basin, showing fault arrays classified by structural style (listric, planar faults, graben and horst) and by fault polarity or displacement directions (synthetic and antithetic faults) (Modified from Chima et al. 2019)

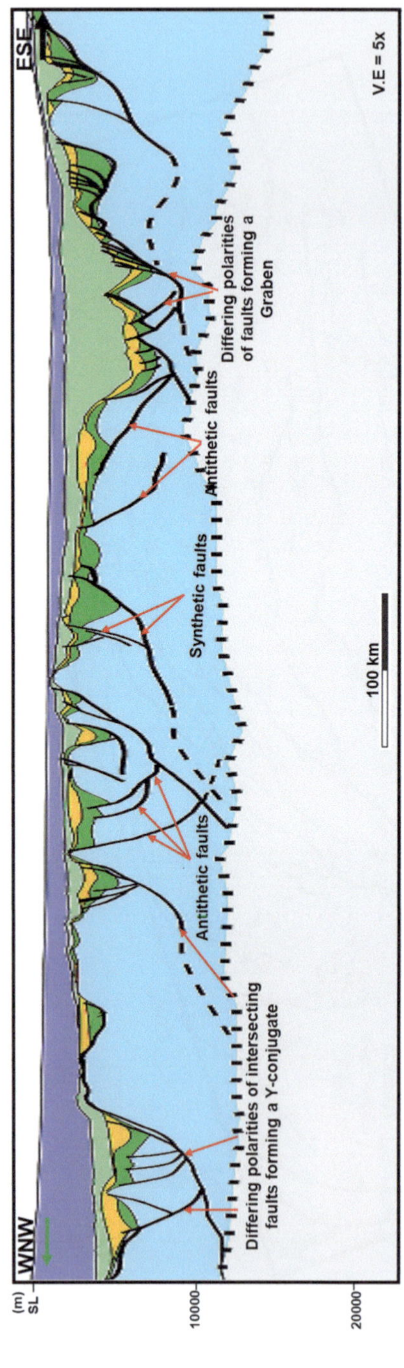

Fig. 9.4 Interpreted regional geoseismic profile of the northern line of the Orphan Basin-Flemish Pass, showing the extensional fault geometry, based on structural styles (listric, planar, graben and Y-conjugate faults) and fault polarity or displacement directions (synthetic and antithetic faults) (Adapted after Cawood et al., 2021a)

Unlike shale- or evaporite-controlled systems such as the Niger Delta and the Levantine Margin, deformation here is closely tied to crustal detachment levels, indicating a more lithosphere-involved extensional regime.

Bounding faults of sub-basins have listric geometries, steepening upward and flattening downward towards detachment horizons. They border isolated depocentres bounded by basement highs, forming narrow, elongated sub-basins 4–5 km wide. This reflects localised strain, with displacement occurring mainly along a few large-offset faults rather than across many small structures.

The Orphan Basin–Flemish Pass features northwest-dipping detachment faults along the southeastern sub-basin margins, extending beneath Jurassic depocentres into deeper-crustal detachments. Fault traces are long and continuous, indicating sustained slip rather than episodic reactivation.

Secondary fault populations comprise synthetic and antithetic normal faults that downthrow hanging-wall blocks towards basin centres. They accommodate differential subsidence within grabens and interact with primary detachments through relay structures and abutments (see Fig. 9.5). Compared with the Levantine Margin and the Niger Delta, these faults are subordinate, primarily fine-tuning accommodation space within a detachment-controlled framework.

The Orphan Basin–Flemish Pass shows how low-angle detachment faults reorganise extensional deformation by focusing slip on large fault surfaces. This creates narrow sub-basins and segmentation, contrasting with shallow gravity-driven systems and revealing diverse extension mechanisms at continental margins.

9.2.4 North Chukchi Basin–Hanna Trough

The North Chukchi Basin–Hanna Trough features basement-involved, steeply dipping extensional faults, unlike detachment-dominated systems (Lothamer, 1994; Minerals Management Service (MMS), 2006). Extension occurs mainly through high-angle normal faults that cut through the basement and sediments, producing a rift with broad fault blocks, significant vertical relief, and limited decoupling (see Fig. 9.6).

Across regional seismic sections, the North Chukchi Basin exhibits north–south–trending normal faults with steep dips that extend through basement reflectors and step stratigraphy downward, forming large fault blocks and asymmetric half-grabens. Unlike margins influenced by evaporite or shale, fault spacing is wider, and deformation concentrates on fewer high-displacement structures rather than on many faults.

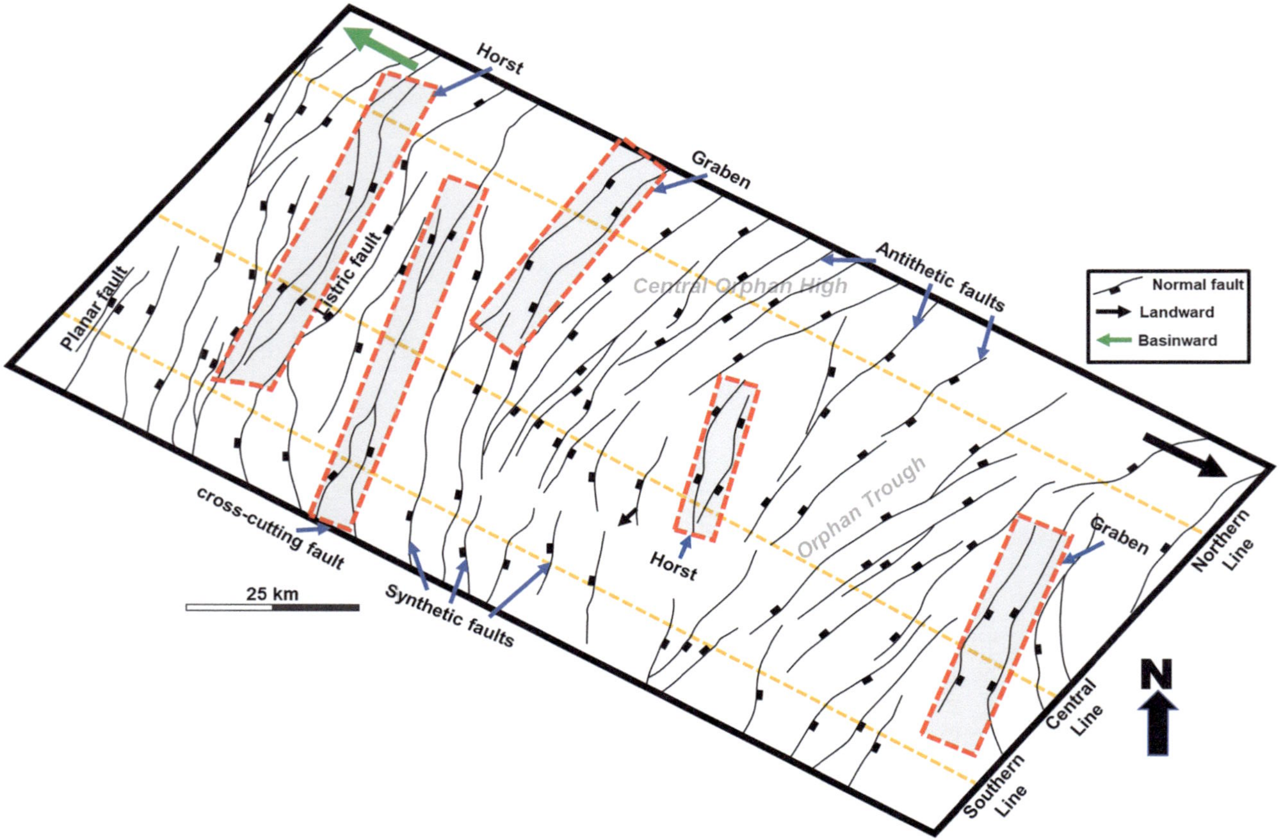

Fig. 9.5 Map view of the Orphan Basin-Flemish Pass, showing the extensional fault geometry based on structural styles (listric, planar, graben, horst and cross-cutting faults) and fault polarity or displacement directions (synthetic and antithetic faults) (Modified from Cawood et al., 2021a)

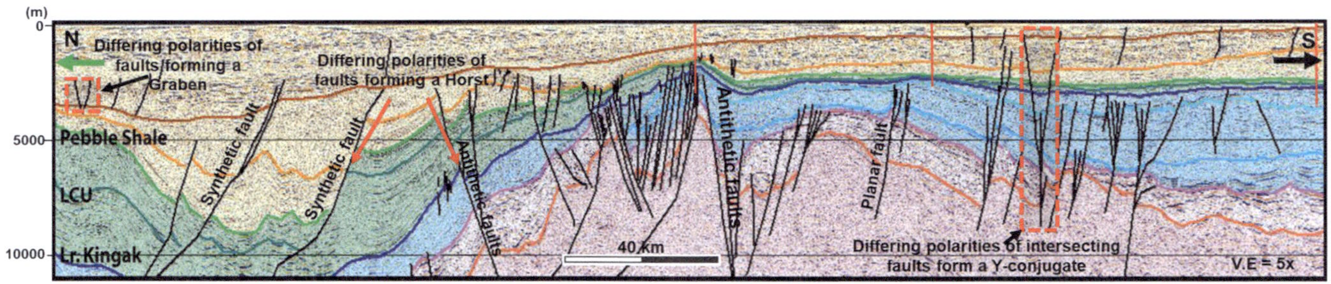

Fig. 9.6 Interpreted regional geoseismic profile of Line 7300 in the North Chukchi Basin, showing the extensional fault geometry, based on structural styles (planar, graben, horst and Y-conjugate faults) and fault polarity or displacement directions (synthetic and antithetic faults) (Adapted from Kumar et al., 2011)

Listric geometries occur where faults flatten into weaker stratigraphic units, forming rollover folds in hanging-wall blocks, primarily in the northern basin. They do not show continuous detachment; instead, they involve local adjustments within a basement-controlled extensional regime.

The Hanna Trough is a distinct sub-domain within the Chukchi system, where faults intersect at ~60°, producing complex graben geometries. These intersections indicate polyphase extension and changing stress directions over time. Structural highs, observed at key wells, are associated with stratigraphic thinning and truncation, whereas nearby growth faults show rapid thickening of Mesozoic and Cretaceous layers.

The North Chukchi Basin–Hanna Trough notably lacks extensive evaporite or shale layers that typically decouple shallow and deep deformation, unlike the Levantine Margin or the Niger Delta. Consequently, shallow fault displacement is closely linked to basement extension, maintaining steep dips throughout much of the stratigraphic column. This mechanical coupling produces a coherent vertical deformation style with fewer secondary faults and less segmentation than on detachment-controlled margins.

In map view, fault traces are long and continuous, indicating sustained slip on principal fault planes (see Fig. 9.7). Minor secondary faults dip basinward and landward, but they are minor within the basin architecture. Overall, the style highlights block faulting and vertical movement, rather than distributed faulting or shallow collapse. Table 9.1 provides a comparative summary of structural styles across selected continental margins.

The North Chukchi Basin–Hanna Trough shows extensional deformation evolving with little mechanical decoupling. Strain localises on steep faults, forming large fault blocks and vertical relief. Detachment horizons control fault geometry, interactions, and basin segmentation.

9.3 Fault Polarities

9.3.1 The Levantine Margin

Fault polarity in the Levantine Margin sector varies and reflects strong decoupling above a weak evaporitic detachment. Both basinward-dipping (synthetic) and landward-dipping (antithetic) normal faults are well developed and often co-occur within the same stratigraphic intervals, particularly in the Plio–Quaternary succession.

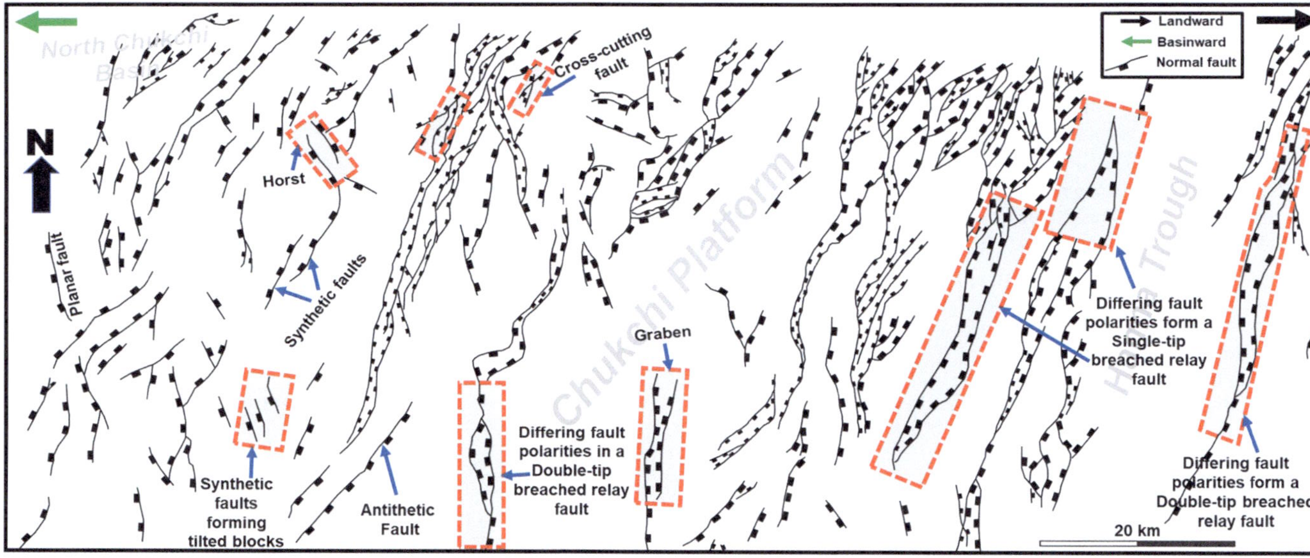

Fig. 9.7 Map view of the North Chukchi Basin and the Hanna Trough, showing the extensional fault geometry based on structural styles (listric, planar, graben, horst, and cross-cutting faults), fault polarity or displacement directions (synthetic and antithetic faults), and fault interactions (single-tip and double-tip relay faults) (Modified from Kumar et al., 2011)

Table 9.1 Comparative summary of structural styles across selected continental margins

Margin	Dominant structural style	Primary fault geometry	Fault spacing and density	Detachment control	Basin architecture expression	Key structural diagnostic
The Levantine Margin	Gravity-influenced, detachment-controlled extensional margin	Planar and listric normal faults; low-angle detachment faults	Closely spaced, high fault density (≈0.25–2 km)	Strong evaporite detachment within Messinian units	Narrow half-grabens and tilted fault blocks; highly segmented margin	V-, X-, and Y-shaped conjugate fault systems above a ductile shear zone
Offshore Western Niger Delta Basin	Shale-detachment–controlled gravity-driven extensional–translational system	Listric growth faults with rollover anticlines; counter-regional faults	Widely spaced in shelf extensional domain; locally dense in translational zones	Strong shale detachment (Akata Formation)	Shelf-scale rollovers transitioning basinward to fold-and-thrust belts	Large-displacement growth faults linked to downslope translation
Orphan Basin–Flemish Pass	Low-angle, crustal-scale detachment-dominated rift system	Shallowly dipping detachment faults; subsidiary listric faults	Widely spaced primary faults; strong fault hierarchy	Mid-crustal to crustal-scale detachments	Narrow (4–5 km) isolated grabens and half-grabens	Basin geometry controlled by long-lived, regionally extensive detachments
North Chukchi Basin–Hanna Trough	Basement-involved extensional rift system	Steeply dipping planar normal faults; minor local listric faults	Widely spaced, low fault density	Minimal to absent regional detachment	Broad fault blocks and asymmetric half-grabens	Strong vertical coupling between basement and cover deformation

Basinward-dipping faults dominate the continental slope, often flattening into Messinian evaporites and accommodating downslope extension through gravitational collapse. Landward-dipping planar faults in shallow layers act as antithetic counterparts, balancing displacement and enabling block rotation in faulted areas (see Figs. 6.7 and 6.9).

This polarity alternation is not random. Fault traces form conjugate patterns, such as V-, X-, and Y-shapes, indicating that polarity is driven by changing stress fields rather than by isolated faults. Antithetic faults often nucleate between major structures, where local stress reorientation and displacement require polarity reversal to maintain kinematic coherence.

The Messinian evaporitic interval facilitates a mixed-polarity system by acting as a ductile shear zone, decoupling shallow faults from deeper levels. This allows polarity to change freely through gravitational sliding and sediment loading, resulting in high variability and short-wavelength alternation in Levantine Margin, unlike margins controlled by basement structure.

Synthetic and antithetic faults coexist, ensuring strain compatibility across segmented margins. Synthetic faults accommodate extension, while antithetic faults absorb rotation and displacement gradients between blocks. This interaction produces balanced extensional systems in which polarity switching facilitates strain partitioning rather than tectonic inversion.

On evaporite-rich margins, fault polarity is dynamic. The offshore Lebanon shows how detachment layers allow quick polarity changes, dense faults, and complex interactions, creating a gravity-driven extension signature over weak layer.

9.3.2 Offshore Western Niger Delta Basin

Fault polarity in the offshore western sector of the Niger Delta Basin is strongly asymmetric and domain-controlled, reflecting the dominant influence of gravity-driven extension above a regionally extensive shale detachment. Unlike the mixed, rapidly alternating polarity patterns observed Levantine Margin, polarity in the Niger Delta is organised into laterally persistent domains in which one dip direction clearly dominates (see Fig. 9.2).

In the shelf extensional domain, basinward-dipping (synthetic) listric growth faults are the principal structures. These faults dip seaward and flatten within overpressured shale units of the Akata Formation, accommodating large displacements and producing prominent rollover anticlines in their hanging walls. Polarity in this domain is therefore highly consistent, with synthetic faults defining both the basin architecture and sediment dispersal pathways.

Downslope, within the translational and transitional zones, polarity becomes more complex yet remains organised. Basinward-dipping faults predominate, whereas landward-dipping faults develop locally, often where sediment load, shale flow, or slope changes perturb the regional stress field. These reversals are not tectonic inversions but kinematic adjustments that maintain displacement in a gravity-driven system.

Fault polarity in the Niger Delta remains asymmetric, with synthetic faults accommodating most extension and antithetic faults acting as secondary structures. Polarity patterns are continuous over tens of kilometres, reflecting mechanical control by the shale detachment and downslope sediment translation (see Fig. 9.3).

From a kinematic assessment, polarity asymmetry in the Niger Delta facilitates efficient strain transfer from the extensional shelf to the compressional toe of the margin. Basinward-dipping faults dominate extension, whereas local antithetic faults accommodate differential block rotation and displacement gradients without disrupting the overall gravitational collapse framework.

On shale-dominated margins, fault polarity shows the direction of gravitational collapse. The offshore western Niger Delta Basin has stable, asymmetric fault patterns due to strong mechanical coupling, with local reversals driven by kinematics, not tectonic regime changes.

9.3.3 Orphan Basin–Flemish Pass

Fault polarity in the Orphan Basin–Flemish Pass is hierarchical and dominated by crustal- to mid–crustal–scale detachment faulting. Unlike gravity-driven margins with frequent polarity changes, this system's polarity is controlled by low-angle detachments, which account for most extension.

The primary polarity is defined by northwest-dipping, shallow detachment faults that control basin shape and form the rift system's backbone by downthrowing hanging-wall blocks towards basin centres (see Fig. 9.4). It remains stable over time and throughout the stratigraphic record, indicating ongoing slip-on long-lived faults.

Secondary fault populations include synthetic and antithetic normal faults, which are smaller and develop along graben margins to accommodate subsidence and block rotation induced by primary detachments. These faults do not disrupt the main polarity framework of the detachment system.

Polarity patterns in the Orphan Basin–Flemish Pass are predictable in cross-section and on maps (see Fig. 9.5). Fault dip directions show limited variability, and polarity reversals are rare and localised. This suggests strong mechanical coupling between upper-crustal deformation and deeper levels, thereby suppressing widespread conjugate fault development.

In detachment-dominated rifts, fault polarity reflects long-lived fault surfaces. The Orphan Basin–Flemish Pass has hierarchical fault systems that produce stable polarity, basin asymmetry, and strain localisation, unlike gravity-driven, shallow-detachment margins.

9.3.4 North Chukchi Basin–Hanna Trough

Fault polarity in the North Chukchi Basin–Hanna Trough is mainly synthetic and vertically consistent, with strong involvement of the basement and limited decoupling (see Fig. 9.6). Unlike detachment margins with alternating polarity, this system has persistent basinward-dipping normal faults that shape the basin from basement to cover.

The dominant polarity is indicated by steeply dipping, basinward-facing faults that form large, asymmetric half-grabens and broad blocks. These faults cut basement reflectors that dip consistently upwards, indicating strong coupling between basement extension and cover deformation. Therefore, polarity remains stable across stratigraphic and strike settings, with little evidence of reversal.

Secondary fault populations comprise small- to mesoscopic planar faults that dip both basinward and landward, often near major synthetic faults. They accommodate minor block rotation and displacement gradients but are spatially limited, subordinate in displacement, and do not alter the basin's primary polarity framework.

Within the Hanna Trough, fault sets often intersect at approximately 60°, creating local complexity in polarity. These intersections indicate polyphase extension and changing stress directions, yet basinward-dipping faults dominate the overall polarity. Variations are confined to specific zones rather than the entire basin (see Fig. 9.7).

Synthetic-polarity dominance enables efficient strain localisation on a few high-displacement faults from a kinematic perspective. The absence of extensive evaporite or shale detachment limits conjugate fault development and polarity switching. Thus, polarity reliably indicates extension direction and basin asymmetry in the North Chukchi Basin–Hanna Trough. Table 9.2 provides a comparative summary of fault polarity across selected continental margins.

The North Chukchi Basin–Hanna Trough shows minimal decoupling, with stable fault polarity linked to basement extension. Polarity better indicates

Table 9.2 Comparative summary of fault polarity across selected continental margins

Margin	Dominant fault polarity	Secondary polarity development	Degree of polarity stability	Primary mechanical control	Kinematic implications	Polarity diagnostic feature
Levantine Margin	Mixed synthetic and antithetic	Frequent alternation; conjugate V-, X-, Y-patterns	Low (high lateral variability)	Evaporite detachment (Messinian units)	Strain partitioning above ductile shear zone; block rotation	Short-wavelength polarity switching above evaporites
Offshore Western Niger Delta Basin	Strongly synthetic (basinward-dipping)	Local counter-regional antithetic faults	Moderate to high (domain-controlled)	Shale detachment (Akata Formation)	Efficient gravity-driven extension with local kinematic adjustment	Laterally persistent polarity domains
Orphan Basin–Flemish Pass	Synthetic, low-angle detachment-controlled	Minor, subordinate antithetic faults	Extremely high (hierarchical)	Crustal-scale detachment faulting	Strong strain localisation onto primary detachments	Long-lived polarity persistence
North Chukchi Basin–Hanna Trough	Predominantly synthetic (steep basinward-dipping)	Local small-scale antithetic planar faults	High (basement-controlled)	Basement-involved extension	Stable basin asymmetry and vertical strain coherence	Polarity continuity from the basement to the cover

extension direction than shallow margins, offering a strong constraint on rift kinematics and basin asymmetry.

9.4 Geometrical and Kinematic Relationships of Fault Interaction

9.4.1 The Levantine Margin

The fault system on the Levantine Margin is highly interactive and developed above ductile evaporites. Faults rarely act in isolation; they form linked systems in which strain is transferred through relay zones, abutments, splays, and cross-cutting features, as shown on maps (see Fig. 9.8).

At the scale of individual fault segments, interaction often occurs through relay fault systems, including convergent, divergent, and synthetic relays. These zones form where adjacent fault tips overlap, creating areas of distributed deformation. Both unbreached relays (where displacement is transferred via folding and minor faulting) and breached relays (with single- and double-tip breaches) exist, indicating progressive fault growth and increasing displacement over time (see Fig. 9.8). Multiple relay styles within the same domain reflect the dynamic evolution of fault linkage rather than a single-stage deformation.

Approaching faults are a key interaction style, propagating with convergent or divergent geometries that reflect local stress and displacement. They often abut, with one fault terminating against the wall of another. These faults may be synthetic or antithetic to the main fault, indicating that interaction depends on kinematic compatibility rather than polarity (see Fig. 9.8).

Splay faults frequently develop at fault tips and along fault planes, branching from main structures into minor faults. These splays distribute strain from the main fault and accommodate local variations in displacement.

Cross-cutting faults support stress perturbations and the overlap of fault activity. They often have long, trailing segments that propagate obliquely across existing fault planes, indicating local stress modifications by earlier faults and resulting in oblique paths rather than straightforward extension.

Fault interaction of the Levantine Margin involves relationships between intersection lines and displacement vectors, with common antithetic–antithetic and antithetic–synthetic interactions (see Fig. 9.9). These shear displacement vectors are often nearly perpendicular to the intersection lines, indicating an extension-dominated strain regime. The bisectors point towards maximum extension, thereby reliably indicating the local strain field (see Fig. 9.9).

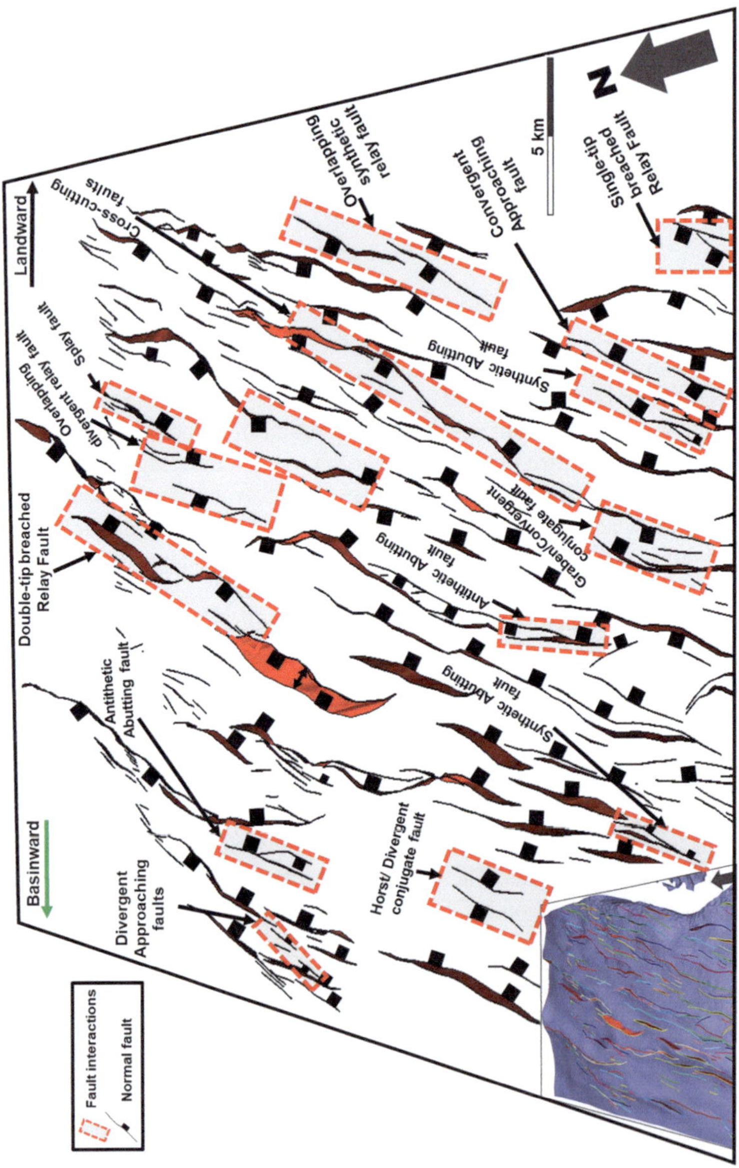

Fig. 9.8 Map view of FM-3 of the Levantine Margin, illustrating the geometric relationships between interacting faults marked in red broken shaded rectangles

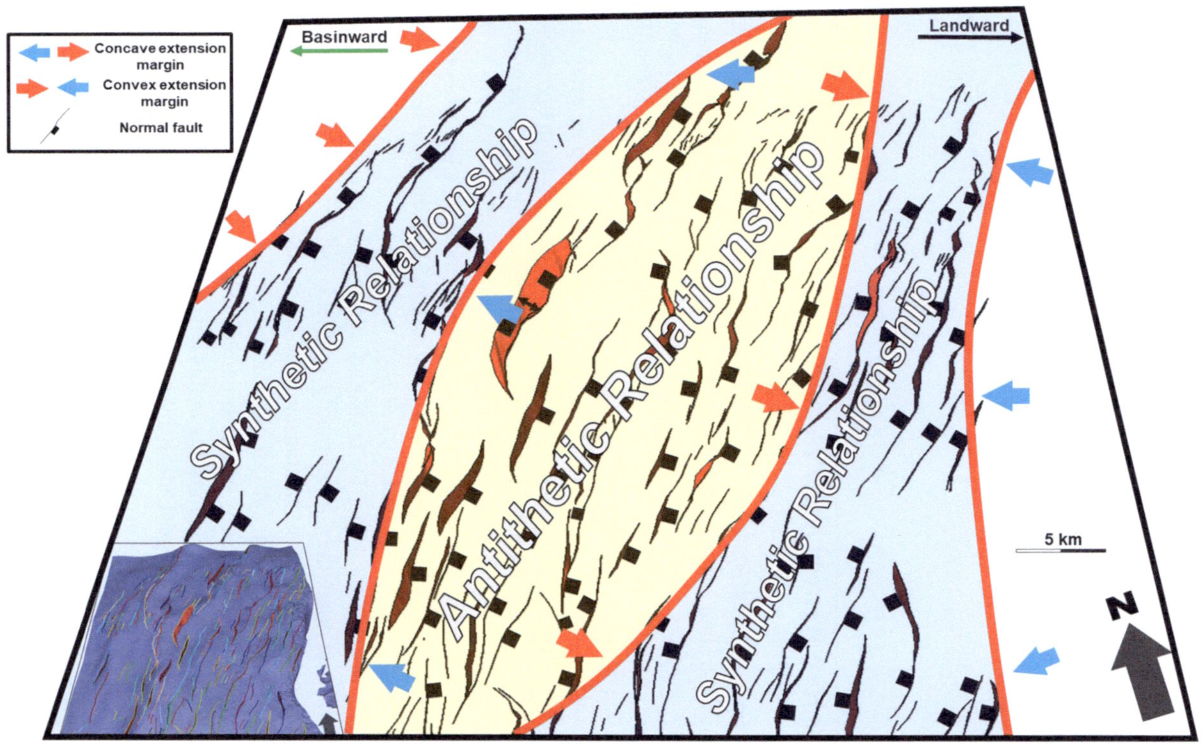

Fig. 9.9 Map view of FM-3 of the Levantine Margin, showing the Antithetic and Synthetic relationships of dominant acute bisectors between interacting faults. The red and orange arrows indicate zones of extension and contraction of normal faults, respectively

Interaction-related kinematics show that polarity switching and fault linkage are essential for displacement compatibility in densely faulted areas (see Fig. 9.9). Synthetic faults accommodate extensional forces, whereas antithetic faults and relay structures absorb block rotation and displacement differences, thereby creating a coherent extensional system despite the complexity.

In evaporite-decoupled margins, fault interaction relies on displacement compatibility, not structural inheritance. The offshore Lebanon shows relay faults, polarity switching, and bisector geometries record extensional history, helping interpret complex fault networks in salt-bearing margins.

9.4.2 Offshore Western Niger Delta Basin

Fault interactions in the offshore western Niger Delta are driven by gravity within a detachment-controlled extensional system, transferring displacement downslope through relay networks. Faults primarily interact within persistent domains, with kinematics governed by sediment loading, shale mobility, and the movement of the sediment wedge above the Akata shale detachment.

Their fault tips interlock. The main interaction involves overlapping relay-fault systems with basinward-dipping listric growth faults that develop from unbreached relay ramps into single-tip breached relays as displacement increases (Evamy et al., 1978; Connor et al., 1998; Corredor et al., 2005; Chima et al., 2022) (see Fig. 9.10). This gradual breaching suggests ongoing fault growth during syn-sedimentary extension, indicating displacement-controlled interaction rather than episodic reactivation (see Fig. 9.10).

Approaching faults in the Niger Delta display strong kinematic organisation. In the shelf extensional zone, they are primarily synthetic and convergent, indicating consistent downslope extension. In translational zones, they may become divergent, reflecting lateral displacement variations and shale flow (see Fig. 9.10). These geometric changes suggest regions where strain redistributes rather than intensifies. There were no abutting fault relationships in this basin.

Fault interactions in the Niger Delta exhibit parallel or subparallel fault intersections and displacement vectors in extensional areas, facilitating the efficient transfer of extension. In more complex transitional zones, intersections become oblique, indicating increased kinematic complexity without altering the extensional regime (see Fig. 9.11).

The orientation of acute bisectors in interacting fault sets indicates extension-dominated strain, with maximum extension roughly perpendicular to the regional fault trend (see Fig. 9.11). Concave and convex fault geometries align with distinct

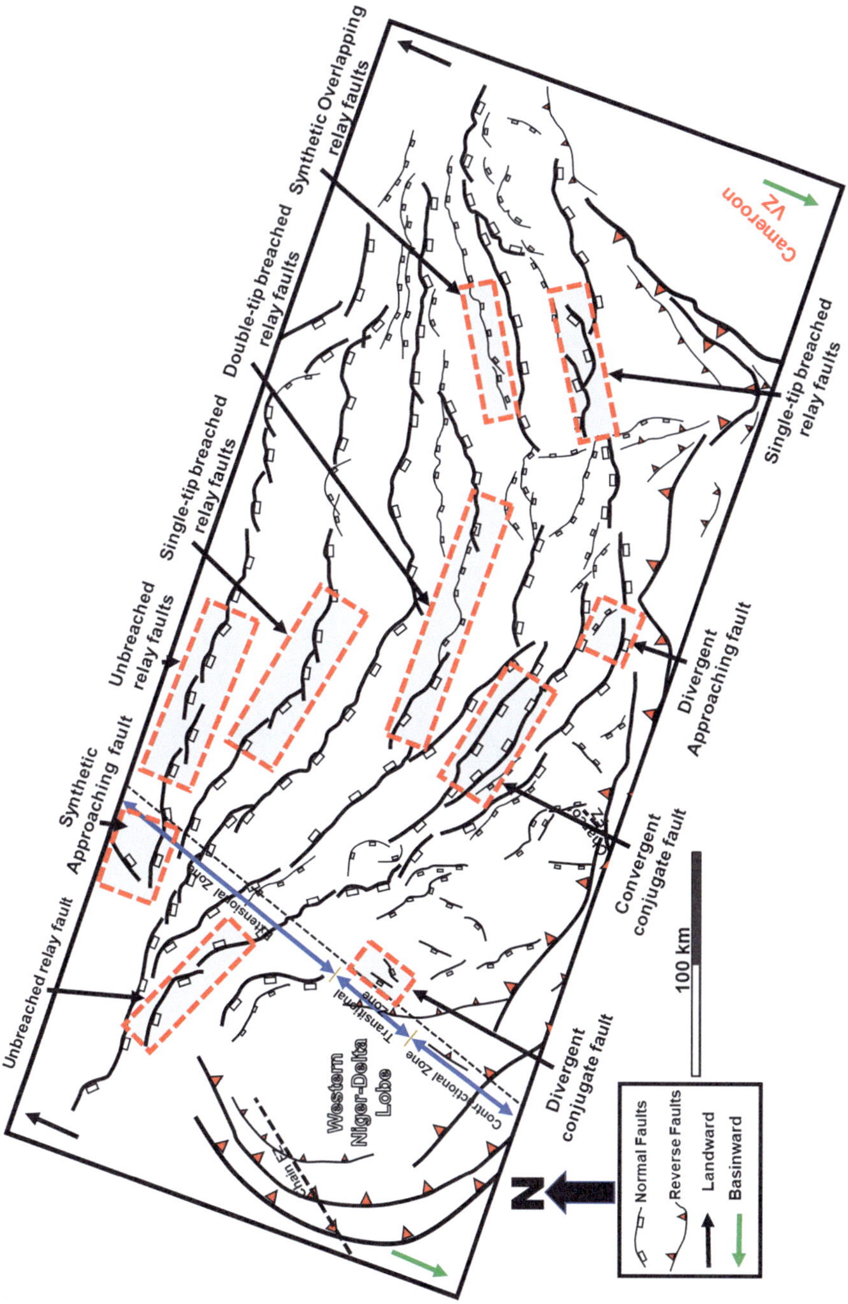

Fig. 9.10 Map illustrating the geometric relationships between interacting faults and linkages highlighted within a red dashed rectangle (Modified after Chima et al., 2019)

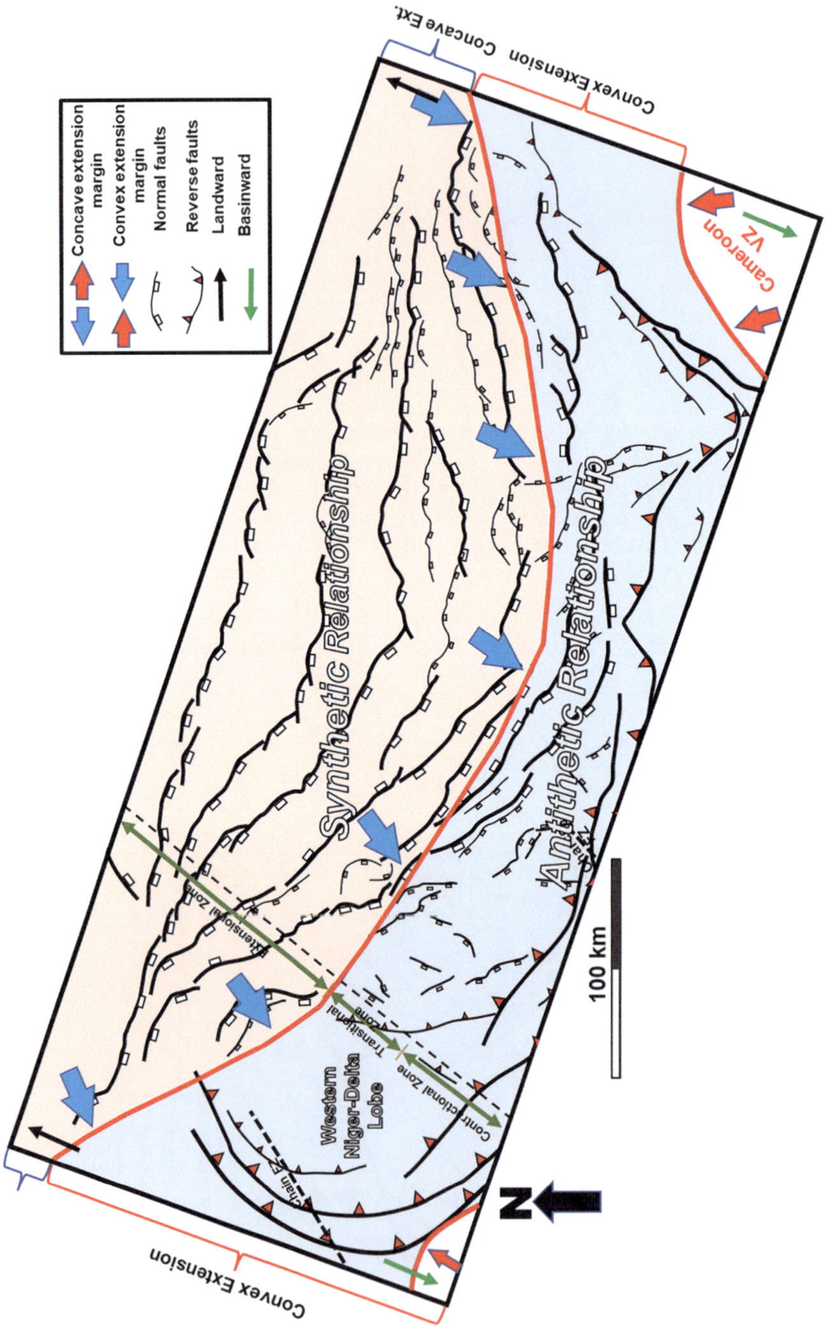

Fig. 9.11 Map view of the offshore Niger Delta Basin, showing the geometric relationships between interacting faults, with linkages highlighted in a red dashed rectangle. (Modified after Chima et al., 2021)

strain zones: concave arrays in extensional areas; convex in tighter, translational or contractional zones.

On shale-detected continental margins, fault interaction kinematics are domain-controlled. The Niger Delta Basin demonstrates how relay systems and synthetic interactions transfer extension efficiently, while local adjustments absorb displacement without disrupting gravity-driven deformation.

9.4.3 Orphan Basin–Flemish Pass

Fault interactions in the Orphan Basin–Flemish Pass form a detachment-dominated extensional system in which strain is transferred primarily along long-lived, low-angle faults. These interactions are fewer than on gravitational margins, yet they are highly organised and efficient, indicating a strong structural and crustal-scale control.

The primary interaction involves relay systems between low-angle detachment faults and subsidiary normal faults, typically breaching at one tip, indicating early linkage and ongoing slip (see Fig. 9.12). Unlike in the Niger Delta or the Levantine Margin, unbreached relay ramps are rare, as extension quickly concentrates on major structures rather than spreading.

Faults in the Orphan Basin–Flemish Pass are often concurrent, propagating together with similar dips and displacements. This indicates stable extension and a strong link between fault segments. Linkage occurs through breaching rather than relay-ramp development, emphasising the dominance of through-going faults (see Fig. 9.12).

Abutting fault relationships typically occur where secondary normal faults terminate against the hanging wall or footwall of primary detachments. These structures are usually synthetic, enabling localised displacement adjustments within narrow grabens (see Fig. 9.12).

Splay faults branch from major detachment surfaces and distribute strain locally without disturbing the main kinematic framework. Their limited presence indicates the efficiency of the detachment system, thereby reducing stress at fault tips (see Fig. 9.12).

In the Orphan Basin–Flemish Pass, kinematic interactions are primarily synthetic–synthetic, with displacement vectors aligned along fault intersections, indicating compatible shear and efficient strain transfer. When the intersection lines are perpendicular, the acute bisectors indicate extension-dominated strain perpendicular to the regional rift axis (see Fig. 9.13).

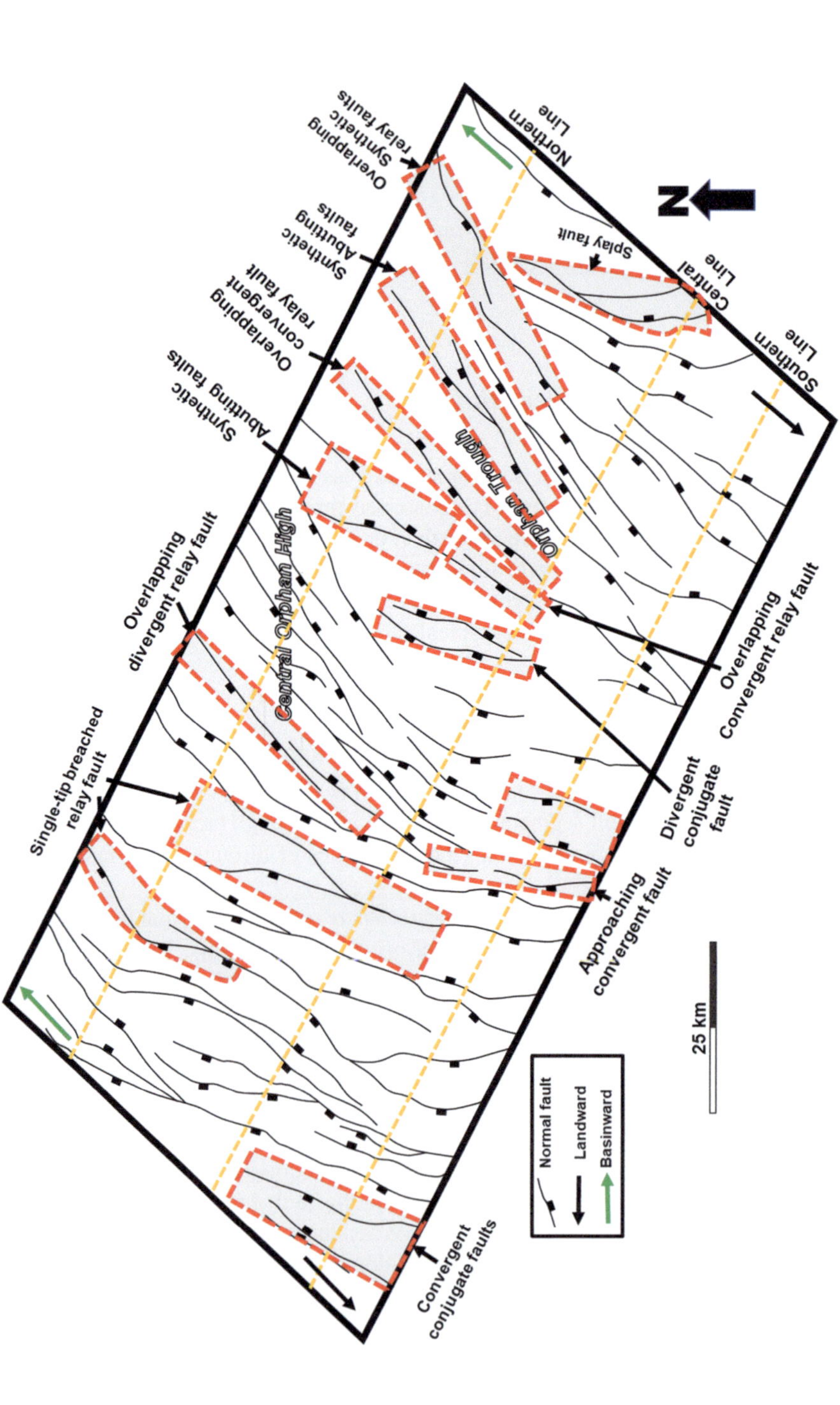

Fig. 9.12 Map view of Orphan Basin-Flemish Pass showing the geometric relationships between interacting faults highlighted in red, broken, shaded rectangles (Modified after Cawood et al., 2021a)

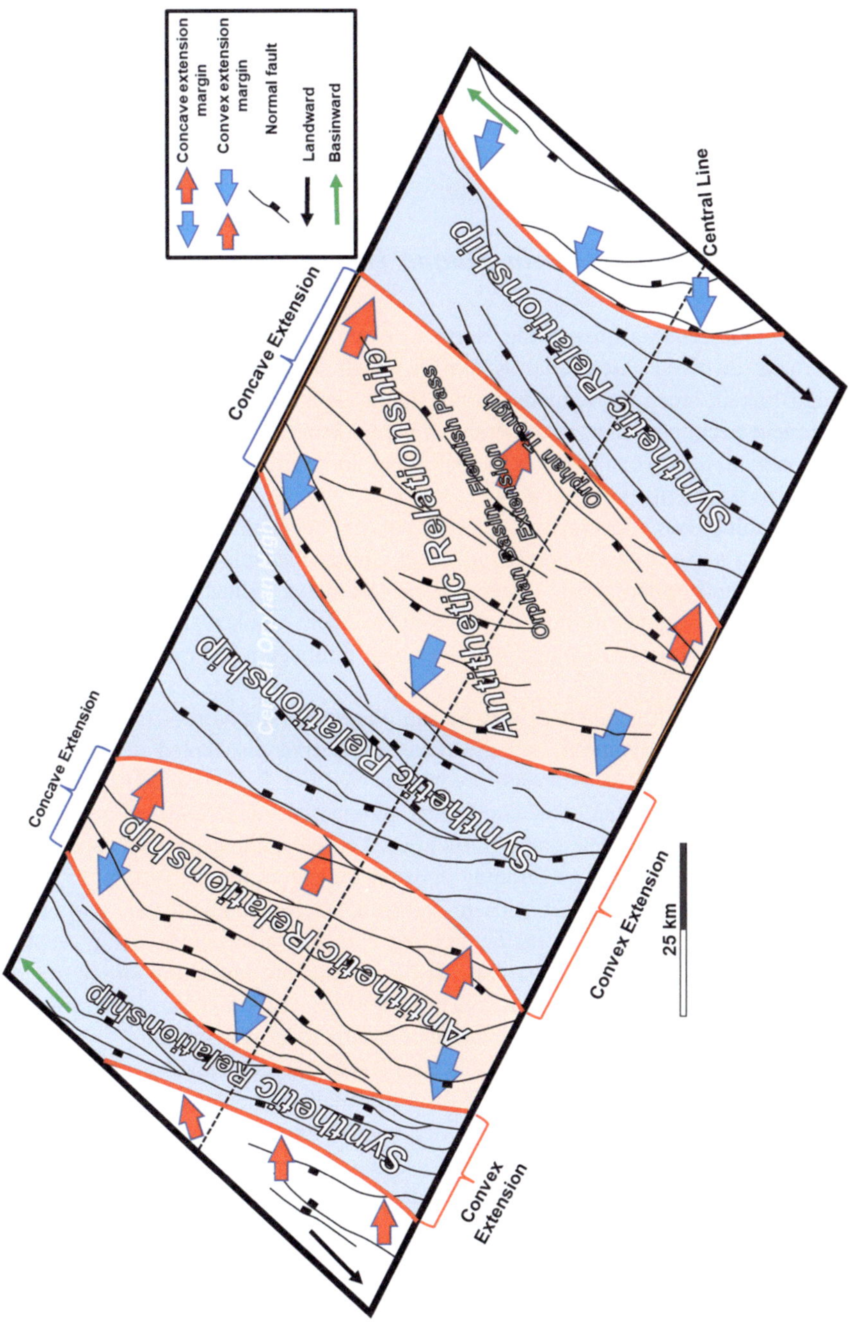

Fig. 9.13 Map view of Orphan Basin-Flemish Pass showing the convex and concave extensions of the Antithetic and Synthetic relationships of dominant acute bisectors between interacting faults. The red arrows indicate the zones of extension (Modified after Cawood et al., 2021a, 2021b)

In detachment-dominated margins, fault interaction favours efficiency over complexity. The Orphan Basin–Flemish Pass shows early fault linkage and stable interactions concentrate strain on long-lived structures, forming narrow sub-basins and predictable patterns.

9.4.4 North Chukchi Basin–Hanna Trough

The kinematic relationships in the North Chukchi Basin–Hanna Trough indicate a basement-involved extensional system in which strain is transmitted primarily through linked, steep, continuous normal faults (see Fig. 9.14). Compared with detachment-decoupled margins, interaction networks are less dense yet more vertically coherent, as major faults cut the basement and transfer displacement upward through the sediments (see Fig. 9.14).

Relay fault systems develop through overlapping, steep normal faults that transfer displacement across relay zones (see Fig. 9.13). Unbreached relays exhibit distributed deformation and folded zones, whereas breached relays indicate fault growth and linkage. This development often results from basement block segmentation, which facilitates the transfer of displacement over long fault segments rather than across many short faults.

Approaching fault geometries vary by sub-domain. In the North Chukchi Basin, faults are generally synthetic and divergent, indicating distributed displacement across multiple strands as extension propagates (see Fig. 9.13). By contrast, faults in the Hanna Trough often show convergent geometries, reflecting localised strain and the formation of discrete graben structures. These differences reveal variability in the stress field and in the partitioning of extension across the rift system.

Abutting relationships occur in basin-and-trough settings. Faults may terminate against major basin-bounding faults, creating complex junctions (see Fig. 9.14). Splay faults, when present, redistribute displacement into branching structures to accommodate local stress changes at fault tips. However, splay faults are less common than at evaporite margins, indicating stronger basement–cover coupling.

Fault interactions in the North Chukchi Basin–Hanna Trough involve various relationships, reflecting main basinward faults with smaller landward ones. The intersection lines and displacement vectors often intersect at oblique angles, indicating extension-dominated strain (see Fig. 9.15). Acute bisectors indicate the principal extension direction and distinguish between distributed extension zones and localised graben formation.

The Hanna Trough features fault junctions intersecting at ~ 60°, indicating non-collinear fault growth and suggesting multiphase extension or variable stress during rifting. Despite local complexity, the system remains kinematically coherent, with

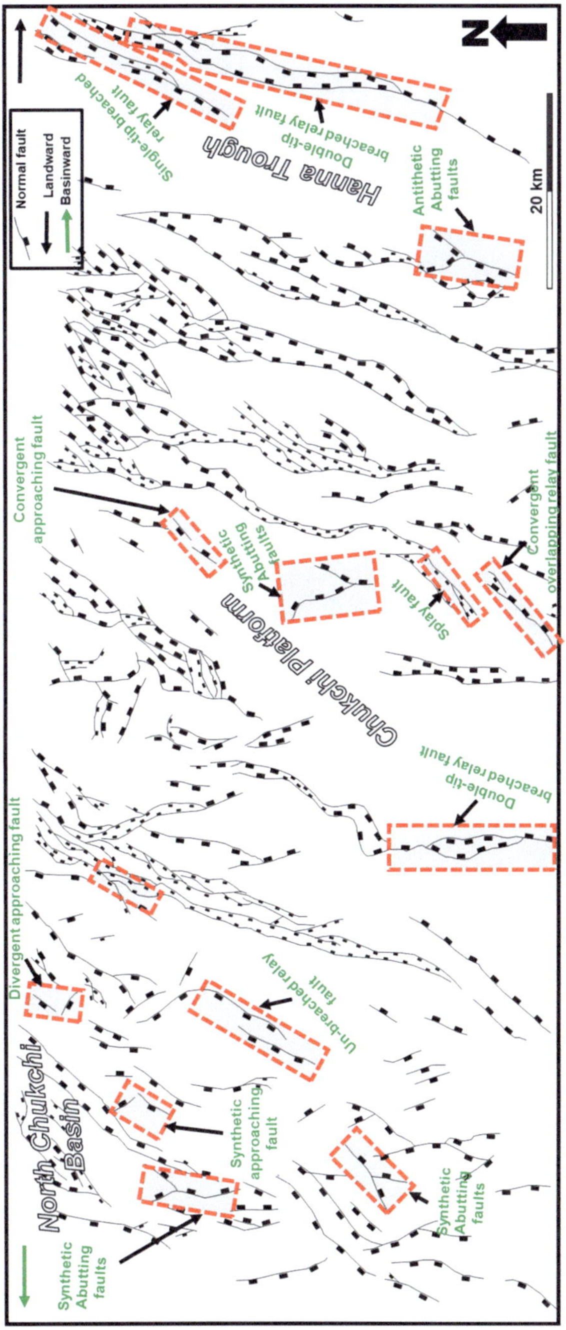

Fig. 9.14 Map view of the North Chukchi Basin and Hanna Trough, illustrating the geometric relationships between interacting faults highlighted in red, broken, shaded rectangles (Modified after Minerals Management Service, 2006)

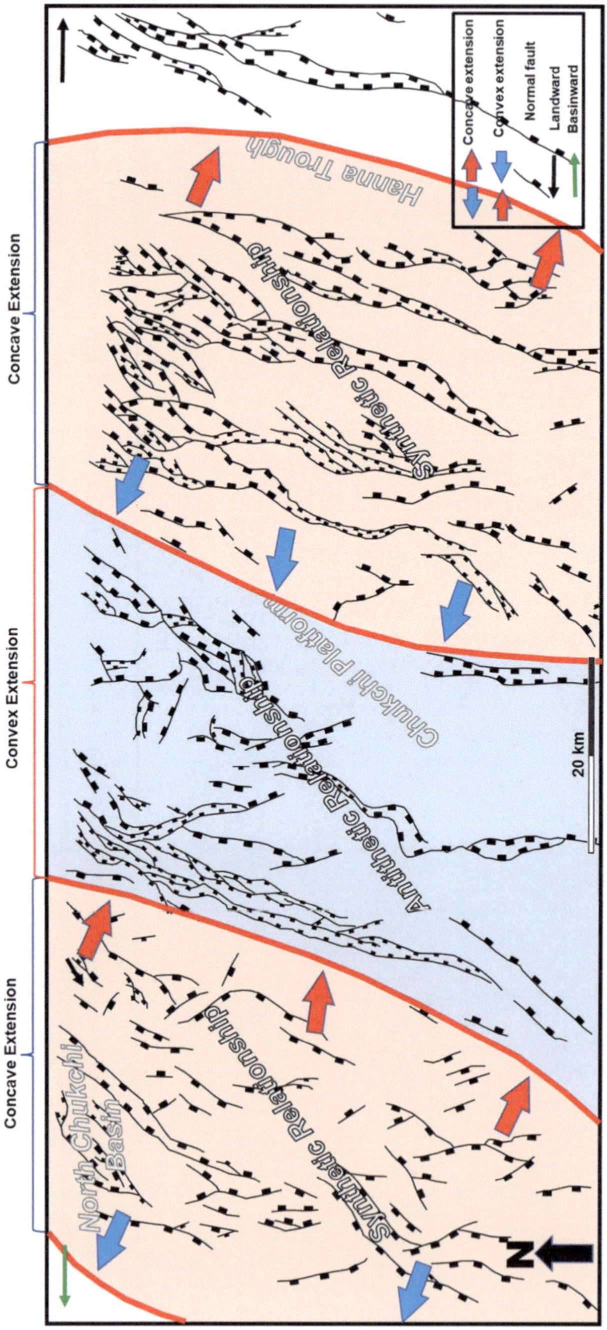

Fig. 9.15 Map view of the North Chukchi Basin and the Hanna Trough, illustrating the convex and concave extensions of the Antithetic and Synthetic relationships of dominant acute bisectors between interacting faults. The red arrows indicate zones of extension (after Minerals Management Service, 2006)

extension accommodated primarily by steep faults that maintain vertical continuity (see Fig. 9.15). Table 9.3 provides a comparative summary of the kinematic relationships of fault interaction across selected continental margins.

In basement-involved rifts, fault interaction depends on crustal-block geometry and segmentation. The North Chukchi Basin–Hanna Trough's relay linkages, fault junctions, and bisectors reveal the extension history and direction amid complex intersections.

9.5 Conclusion and Implications

This chapter integrates structural styles, fault polarity, and kinematic relationships of fault interactions to compare the Levantine Margin with three contrasting continental margins. Taken together, these descriptors provide a process-based explanation for why rifted margins with similar extensional settings develop markedly different fault networks and basin architectures.

9.5.1 *Conclusion*

Structural styles define the architectural template for deformation. Shallow, weak detachments promote dense, segmented fault networks, whereas deep or absent detachments favour early fault linkage and hierarchical systems. Across the examined margins, structural style scales with detachment depth and rheology, from distributed faulting above weak horizons to vertically coherent deformation in basement-involved rifts.

Fault polarity records the degree of mechanical decoupling. Mixed or rapidly alternating polarity indicates high kinematic freedom above weak detachments, whereas stable, predictable polarity indicates strong coupling between the basement and the cover. Polarity patterns, therefore, serve as a practical diagnostic for identifying the influence of detachment and strain asymmetry in extensional systems.

Kinematic relationships in fault interactions reveal the growth of fault networks. Relay ramps, abutting and splay faults, cross-cutting links, and bisector orientations document how displacement is transferred and localised over time. Distributed systems show adaptive, relay-rich kinematics, whereas localised systems exhibit early linkage and strain transfer along a few major faults.

Table 9.3 Comparative summary of kinematic relationships of fault interaction across selected continental margins

Margin	Dominant fault interaction style	Relay system characteristics	Approaching fault geometry	Abutting and splay fault development	Intersection–displacement relationship	Kinematic signature
The Levantine Margin	Highly interactive, displacement-balanced fault networks	Both unbreached and breached relays (single- and double-tip); diverse relay geometries	Convergent and divergent; synthetic and antithetic	Abundant abutting and splay faults; frequent cross-cutting	Displacement vectors are commonly perpendicular to intersection lines	Acute bisectors indicating extension-dominated strain above evaporite detachment
Offshore Western Niger Delta Basin	Domain-organised, gravity-driven relay interaction	Predominantly overlapping relays evolving to single-tip breached ramps	Synthetic and convergent in the extensional domain; divergent in the translational zones	Abutting faults are common but domain-restricted; limited splaying	Intersection lines are commonly parallel or sub-parallel to displacement vectors	Synthetic–synthetic interactions reflecting efficient downslope strain transfer
Orphan Basin–Flemish Pass	Hierarchical, detachment-dominated interaction	Early-linked, mainly single-tip breached relays; few unbreached ramps	Concurrent and synthetic; stable interaction geometry	Limited abutting and splay development	Displacement vectors are parallel primarily to the intersection lines	Stable synthetic–synthetic kinematics with strong strain localisation
North Chukchi Basin–Hanna Trough	Basement-rooted, vertically coherent interaction	Both unbreached and breached relays between long fault segments	Synthetic–divergent in basin; convergent in Hanna Trough	Abutting and minor splay faults are present but sparse	Intersection lines are oblique to displacement vectors	Mixed kinematic interactions with acute bisectors recording extension direction

9.5.2 *Integrative Implications for Rift Evolution*

When combined, the three descriptors define a coherent continuum of rift behaviour:

- Distributed end-member: shallow detachment–controlled margins with high fault density, mixed polarity, and complex interaction networks.
- Translational end-member: shale-detached margins with domain-organised structure, asymmetric polarity, and efficient relay systems.
- Localised end-member: crustal- or basement-involved margins with stable polarity, early fault linkage, and vertically coherent kinematics.

Importantly, margins are not static. Changes in sedimentation rate, detachment efficiency, or extension magnitude can shift a system along this continuum, thereby altering fault density, polarity organisation, and interaction style over time.

9.5.3 *Practical Implications*

Structural Interpretation: Integrating geometry, polarity, and kinematics reduces ambiguity by distinguishing kinematic polarity reversal from tectonic inversion and by identifying true linkage from apparent segmentation.

Basin analysis and resource assessment: Fault interaction style influences compartmentalisation, fluid pathways, and seal integrity. Distributed relay networks may enhance connectivity, whereas early-linked fault systems may promote large, isolated compartments.

Comparative Tectonics: The framework developed here enables the transfer of insights between margins by matching mechanical conditions rather than relying on superficial geometric similarity.

Learning Objectives

After completing this chapter, readers will be able to:

1. Compare fault architectures across different continental margins.
2. Evaluate similarities and contrasts between salt-influenced and basement-involved rifts.
3. Assess the role of tectonic inheritance in fault development.
4. Transfer interpretation workflows between geological settings.
5. Critically evaluate the generality of fault growth models.

References

Brun, J. P., & Choukroune, P. (1983). Normal faulting, block tilting and décollement in a stretched crust. *Tectonics, 2*(4), 345–356.

Buck, W. R. (1991). Modes of continental lithospheric extension. *Journal of Geophysical Research: Solid Earth, 96*(B12), 20161–20178.
Cawood, A. J., Ferrill, D. A., Morris, A. P., Norris, D., McCallum, D., Gillis, E., & Smart, K. J. (2021a). Tectonostratigraphic evolution of the Orphan Basin and Flemish Pass region—Part 1: Results from coupled kinematic restoration and crustal area balancing. *Marine and Petroleum Geology, 128*, Article 105042.
Cawood, A. J., Ferrill, D. A., Morris, A. P., Norris, D., McCallum, D., Gillis, E., & Smart, K. J. (2021b). Tectonostratigraphic evolution of the Orphan Basin and Flemish Pass region—Part 2: Regional structural development and lateral variations in rifting style. *Marine and Petroleum Geology, 133*, Article 105219.
Chima, K. I., Do Couto, D., Leroux, E., Gardin, S., Hoggmasacall, N., Rabineau, M., Granjean, D., & Gorini, C. (2019). Seismic stratigraphy and depositional architecture of Neogene intraslope basins, offshore western Niger Delta. *Marine and Petroleum Geology, 109*, 449–468.
Chima, K. I., Granjeon, D., Couto, D. D., Leroux, E., Gorini, C., Rabineau, M., Letouzey, J., Hoggmascall, N., & Glukstad, M.-M. (2022). Tectono-stratigraphic evolution of the offshore western Niger Delta from the Cretaceous to present: Implications of delta dynamics and paleotopography on gravity-driven deformation. *Basin Research, 34*, 25–49. https://doi.org/10.1111/bre.12609
Connors, C. D., Denson, D. B., Kristiansen, G., & Angstadt, D. M. (1998). Compressive anticlines of the mid-outer slope, central Niger Delta (abs). *AAPG Bulletin, 82*, 1903.
Corredor, F., Shaw, J., & Bilotti, F. (2005). Structural styles in the deepwater fold and thrust belts of the Niger Delta. *AAPG Bulletin, 89*, 753–780.
Damuth, J. (1994). Neogene gravity tectonics and depositional processes on the deep Niger Delta continental margin. *Marine and Petroleum Geology, 11*, 320–346.
Evamy, B. D., Haremboure, J., Kamerling, P., Knapp, W. A., Molloy, F. A., & Rowlands, H. (1978). Hydrocarbon habitat of the Tertiary Niger Delta. *AAPG Bulletin, 62*, 1–39.
Garfunkel, Z., & Almagor, G. (1985). Geology and structure of the continental Margin off northern Israel and the adjacent part of the Levantine Basin. *Marine Geology, 62*, 105–131.
Graue, K. (2000). Mud volcanoes in deepwater Nigeria. *Marine and Petroleum Geology, 17*(8), 959–974.
Kumar, N., Granath, J.W., Emmet, P.A. Helwig, J.A. and Dinkelman, M.G. 2011. Stratigraphic and tectonic framework of the U.S. Chukchi Shelf: exploration insights from a new regional deep-seismic reflection survey. From: Spencer, A. M., Embry, A. F., Gautier, D. L., Stoupakova, A. and Sørensen, K. (eds) *Arctic Petroleum Geology*. Geological Society, London, Memoirs, 35, pp. 501–508.
Lothamer, R. T. (1994). Early tertiary wrench faulting in the North Chukchi basin, Chukchi Sea, Alaska. In D. K. Thurston, & K. Fujita (eds.) (1992) *Proceedings International Conference on Arctic Margins* (pp. 252–256). U.S. Minerals Management Service, Anchorage, AK, OCS Studies, MMS 94-0040.
Minerals Management Service (MMS). (2006). Chukchi sea planning area (Alaska)—province summary, oil and gas assessment (pp. 1–36).
Morley, C. K., & Guerin, G. (1996). Comparison of gravity-driven deformation styles and behaviour associated with mobile shales and salt. *Tectonics, 15*(6), 1154–1170.
Nelson, M. A. (2013). 3D geometry and kinematics of non-colinear fault intersections (PhD thesis). Cardiff University, U.K.
Osmond, J. C. (1960). Tectonic history of the Basin and Range province in Utah and Nevada. *Mining Engineering, 12*(3), 251–265.
Park, R. G. (1993). *Geological structures and moving plates*. Chapman & Hall.
Rouby, D., Nalpas, T., Jermannaud, P., Robin, C., Guillocheau, F., & Raillard, S. (2011). Gravity-driven deformation controlled by the migration of the delta front: The Plio-Pleistocene of the Eastern Niger Delta. *Tectonophysics, 513*, 54–67.
Short, K., & Stauble, A. (1967). Outline of geology of Niger Delta. *AAPG Bulletin, 51*, 761–779.

Chapter 10
Comparative Studies of the Levantine Margin and Analogue Sandbox Model Experiments

Abstract This chapter compares seismic-scale fault systems on the Levantine Margin with analogue sandbox model experiments, focusing on structural styles, fault polarities, and the kinematic relationships of fault interactions. Interpreted 2D and 3D seismic data are combined with established analogue models to investigate how extensional faults nucleate, develop, interact, and connect over time. The analysis highlights recurring structural features, including planar and listric normal faults, rotated fault blocks, graben–horst architectures, detachment surfaces, and conjugate fault systems. Map-scale fault analysis reveals segmented fault growth, elliptical fault tips, relay development, and progressive linkage driven by displacement accumulation and extension-dominated strain. Kinematic assessment shows that the Levantine Margin maintains a wide range of interaction patterns, from antithetic–synthetic to synthetic–synthetic relationships, reflecting fault maturity, mechanical stratigraphy, and stress heterogeneity. These features are compared with analogue oblique-rift experiments, which replicate the simplified end-member kinematics of mature fault systems. By integrating seismic observations with experimentally constrained deformation, this chapter demonstrates that analogue models offer an essential mechanical framework, while seismic data capture the full complexity of natural fault networks.

Keywords Extensional tectonics · Graben–horst architecture · Conjugate faults · Oblique rifting · Analogue modelling · Kinematic analysis

10.1 Introduction

This chapter presents a framework for comparing the Levantine Margin with classic sandbox experiments, focusing on structural styles, fault polarities, and kinematic relationships. Integrating 2D and 3D seismic observations with experimental fault models provides a physically grounded understanding of fault nucleation, propagation, interaction, and linkage at continental margins.

This approach links natural and experimental systems by comparing seismic observations with scaled analogue models that have known boundary conditions and

K. Chiadikobi, *Advances in Horizon Mapping and Fault Interpretation*,
https://doi.org/10.1007/978-3-032-23917-4_10

deformation histories. McClay and Ellis's (1987) models and those of McClay et al. (2002) serve as references because they reproduce fault growth, including planar and listric normal faults, rotated blocks, graben–horst structures, detachment surfaces, and conjugate faults. They also document fault-interaction processes, such as relay development, linkage, breaching, and cross-cutting, which are vital for understanding the structural evolution of the Levantine Margin.

This chapter examines fault polarity (synthetic versus antithetic) and interaction kinematics, which vary spatially and temporally in natural systems and analogue experiments. Faults display basinward- and landward-dipping geometries, antithetic–synthetic pairs, and mixed patterns influenced by stratigraphy, inherited structures, and rheology. These patterns are compared with sandbox models in which deformation depends on extension direction, basement configuration, and material uniformity, and are scaled vertically to match seismic interpretation scales.

The chapter guides the reader from geometry to process by comparing seismic structural styles with experimental models. It then examines fault polarities in terms of symmetry, dip direction, and displacement within extensional systems. Finally, it examines how displacement, geometry, and fault orientation influence linkage styles, including relay ramps, breached relays, and splay faults.

Treating analogue experiments as mechanical analogues, this chapter highlights how they both converge with and diverge from natural fault systems. This synthesis enhances seismic interpretation, clarifies the evolution of fault networks, and supports detailed analyses in subsequent sections.

The analogue sandbox model of McClay and Ellis (1987) and McClay et al. (2002) was a deliberate choice that requires proper justification in the broader landscape of available models. While various sandbox models of extensional faulting have been developed over the years, the rationale behind opting for these specific models is complex and can be explained as follows:

- The sandbox models by McClay and Ellis (1987) and McClay et al. (2002) are recognised for their foundational role in understanding fault interactions and linkages within extensional tectonic settings. Their pioneering nature has established them as key references in the field, as benchmarks against which newer models can be compared.
- These sandbox models have laid the groundwork for interpreting fault behaviours, interactions, and complexities. The comprehensive nature of these models, encompassing a range of fault geometries and interactions, makes them suitable for comparison with the Levantine Margin.
- Over the years, the models by McClay and Ellis (1987) and McClay et al. (2002) have undergone validation and comparison against real-world geological observations and seismic data. Their ability to reproduce features observed in natural systems enhances their credibility and makes them valuable references for seismic interpretation studies.

10.2 Comparing Structural Styles of Seismic Profiles of the Levantine Margin and Analogue Sandbox Model

This section compares the structural styles observed on selected 2D seismic profiles from the Levantine Margin with those developed in classic analogue sandbox experiments. The comparison is strictly geometric and kinematic, focusing on how fault architectures evolve under extensional strain. Emphasis is placed on fault geometry, fault-block organisation, and the spatial arrangement of fault systems, thereby providing a direct bridge between seismic interpretation and experimentally constrained deformation processes.

The aim is to identify recurring structural patterns that reflect shared mechanical controls on fault nucleation, growth, and interaction, rather than to establish a one-to-one similarity between natural and analogue systems. These patterns help interpret the evolution of fault systems along the Levantine margin.

10.2.1 Levantine Margin versus the Analogue Model of McClay and Ellis (1987)

The analogue sandbox model by McClay and Ellis (1987) provides a key experimental reference for analysing extensional fault systems. Their model shows the progressive development of normal faults above a basement discontinuity, capturing the transition from early fault nucleation to mature graben formation (see Fig. 10.2). When compared with seismic profiles (notably Lines 22,000, 21,400, and 21,920, see Fig. 10.1), several main similarities in structural style are evident.

The structural styles in the localised region line 22,000 are steeply dipping, apparently planar boundary faults (e.g., Fault 2; Fig. 10.3a), resembling those in the McClay and Ellis (1987) analogue model (Fault 2; Fig. 10.3b). Both display a graben in which new faults develop in the graben's hanging wall.

Apparent planar faults (e.g., Faults 1 and 14; Fig. 10.4a) and conjugate faults (e.g., V-shaped Faults 5 and 11; Y-shaped Faults 3 and 13; Fig. 10.4a) are present. Faults within the graben structure of the analogue model are mainly planar (e.g., Faults 1, 3, 4, 5, 6; Fig. 10.4b). Some faults nucleate within boundary faults, with evidence of rotated fault blocks (e.g., Faults 5, 10, 4, 3, 9; Fig. 10.4a). The western margin of the regional section of Line 22,000 displays steep, evenly spaced faults (e.g., Faults 3, 4, and 1; see Fig. 10.4a). These faults resemble the analogue sandbox model by McClay and Ellis (1987) (e.g., Faults 2 and 5; Fig. 10.4b). A slight rotation of fault blocks is also observed, similar to that observed in analogue experiments (Faults 1, 3, 4, and 6; Fig. 10.4b).

The localised section of Line 21,400 includes steeply dipping boundary faults (Fault 1; see Fig. 10.5a) that cut through Plio-Quaternary and Top Messinian strata with different polarities (e.g., right-lateral dip basinward, left-lateral dip landward).

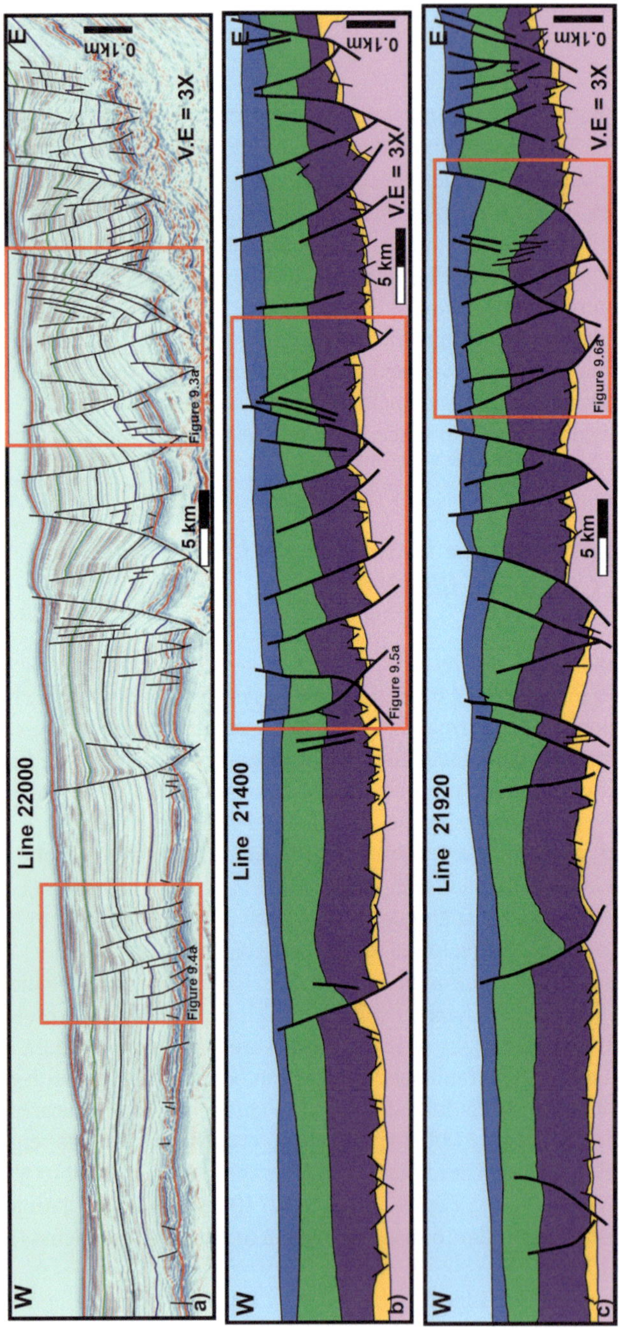

Fig. 10.1 Interpreted seismic profiles showing **a** Line 22,000, **b** Line 21,400 and **c** Line 21,920. Inserts indicate the localised areas of the sections

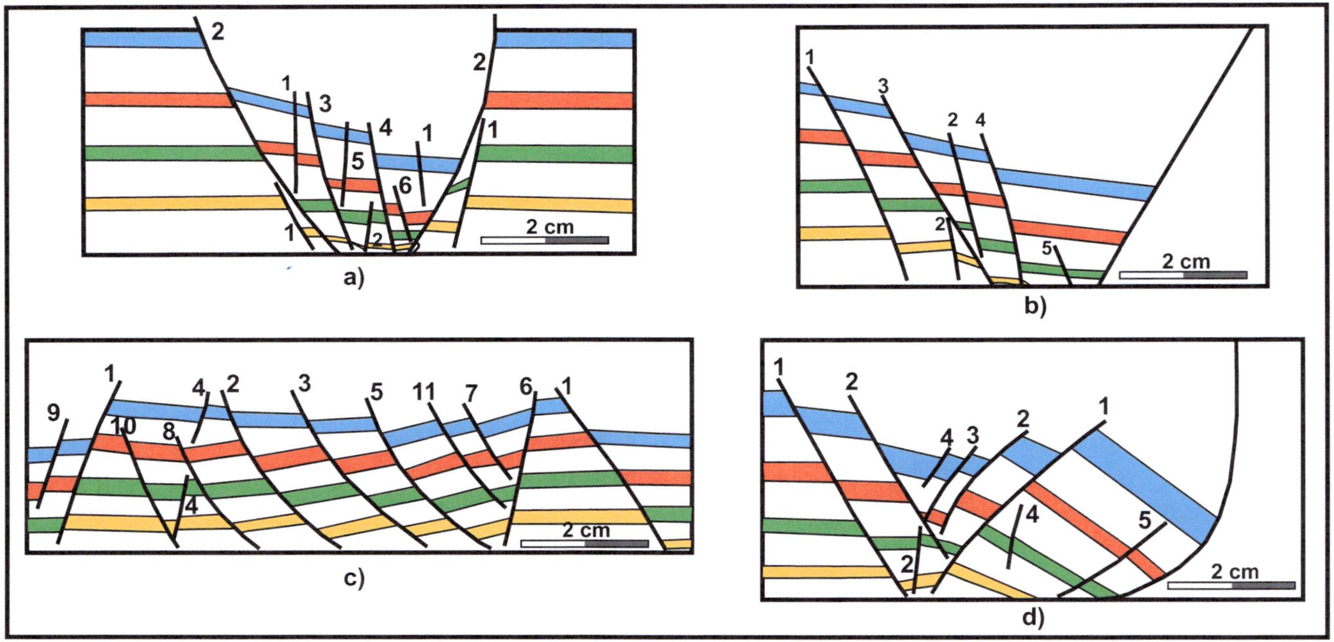

Fig. 10.2 Fault-sequence diagram derived from analogue sandbox experiments of McClay and Ellis (1987) showing **a** dominant nucleation of faults into the hanging wall of the Graben. Minor footwall collapse has truncated the high-angled fault at 25% extension, **b** dominant nucleation of faults into the hanging wall at 33% extension, **c** the development of evenly spaced positive Listric faults with rotated layering between the fault surfaces at 50% extension, **d** planar and negative Listric-fault development in a crestal collapse graben at 33% extension

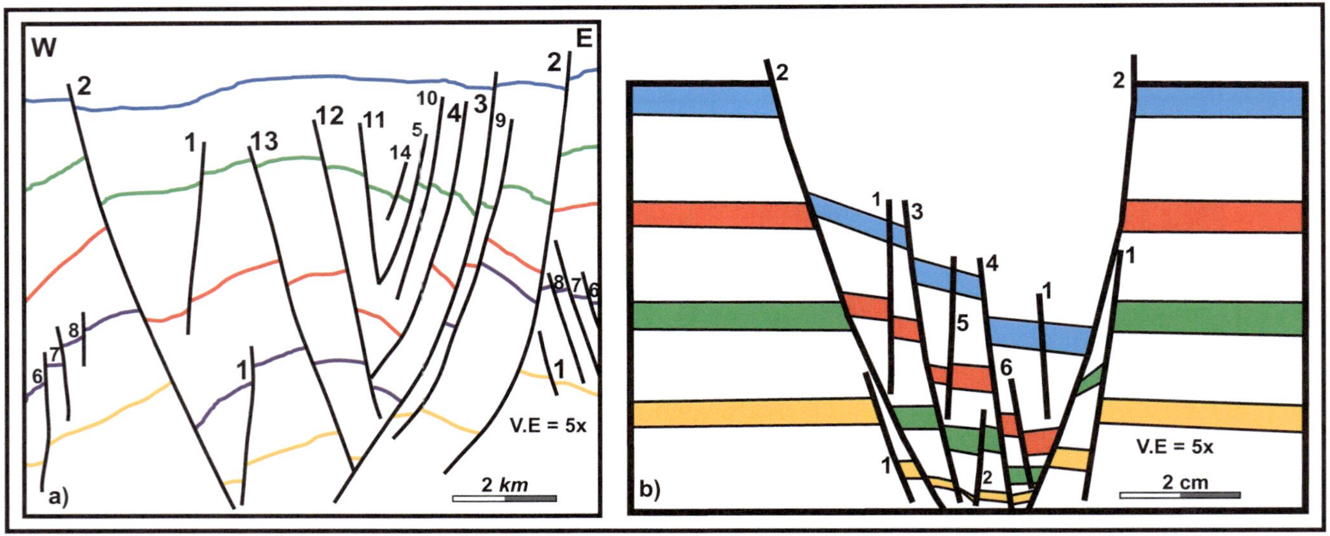

Fig. 10.3 **a** Localised seismic profiles showing structural styles of Line 22,000. Fault 2 is a boundary fault; Faults 1 and 14 are planar; Faults 5 and 11 are V-shaped; Faults 3 and 13 are Y-shaped; Faults 9, 3, 4, 10, 5 are left-stepping rotated blocks; Faults 13, 12, 11 are right-stepping blocks. **b** Analogue model by McClay and Ellis (1987) showing structural styles from extension over a linear basement dislocation. Fault 2 is a boundary; Faults 1, 3, 4, 5, 6 are planar

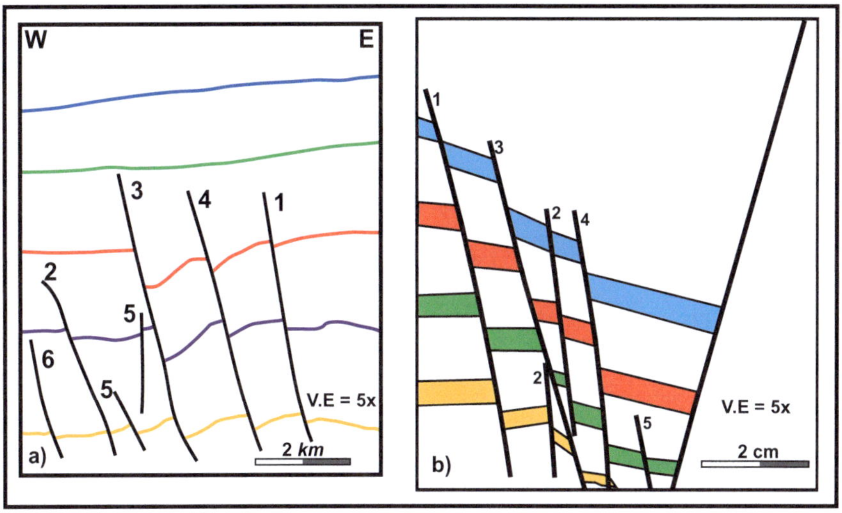

Fig. 10.4 **a** Localised seismic profile showing structural styles of Line 22,000 with faults 1, 3, 4, 5, and 6 as planar faults. **b** Analogue model by McClay & Ellis, 1987, illustrating structures under extension controlled by a planar fault (Faults 2 and 5 as planar faults; Faults 1, 3, 4, and 6 as right-stepping rotated fault blocks)

These boundary faults are similar to those seen in analogue sandbox experiments (Fault 1; Fig. 10.5b). Apparent listric faults within the boundary faults mostly dip landward, are evenly spaced, and show rotated layers between fault surfaces. Fault-block rotation was observed (Faults 4, 5, 2, and 3; Fig. 10.5a) and in the analogue model (Faults 10, 8, 2, 3, 5, 11, and 7; Fig. 10.5b).

The localised section of Line 21,920 displays an apparent listric fault on the right (Fault 2) and a planar fault on the left (Fault 1), both crossing the Plio-Quaternary and Top Messinian layers (see Fig. 10.6a). These boundary faults are similar to those observed in analogue sandbox experiments (Fig. 10.6b). Within these faults, X conjugate faults (Faults 10 and 11; 4 and 10), rotated fault blocks (Faults 6 and 7), and planar faults (Faults 3, 5, 8, and 9) were identified in the Levantine Margin (see Fig. 10.6a), whereas Y-shaped conjugate faults (2 and 3) and rotated blocks (Faults 7, 8, 3, and 4) are seen in the analogue sandbox experiments (see Fig. 10.6b).

The comparison indicates that the Levantine Margin shares key features with the McClay and Ellis (1987) model, including boundary-controlled graben formation, fault nucleation, rotated blocks, and fault curvature. These similarities confirm that the seismic data aligns with extensional models and support using analogue sandbox experiments to interpret fault architecture.

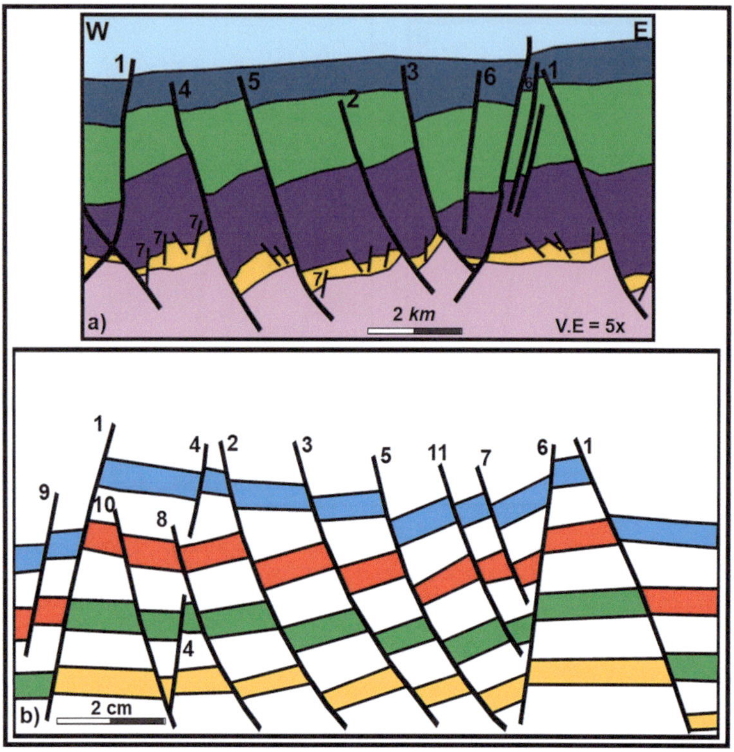

Fig. 10.5 **a** Localised seismic profiles show different structural styles on X-line 21,400. Fault 1 marks boundary faults with opposite polarities; one dips landward, the other basinward. Faults 4, 5, 2, and 3 are right-stepping rotated blocks; Faults 6 and 7 are planar. **b** McClay and Ellis's analogue model (1987) illustrates structures from extension above a stretched basement. Fault 1 is a boundary; Faults 10, 8, 2, 3, 5, 11, and 7 are right-stepping rotated blocks; Faults 4 and 9 are planar

10.2.2 Levantine Margin versus the Analogue Model of McClay et al. (2002)

This compares the structural styles from a regional seismic profile of the Levantine Margin (see Fig. 10.7a) with the analogue (sandbox) model of McClay et al. (2002) for the oblique rift basin experiment at 60° and 45° (see Fig. 10.7b, c) relative to the direction of extension. Unlike McClay and Ellis's (1987) orthogonal extension, these experiments focus on fault development under oblique extension, which is relevant for continental margins with strain partitioning parallel and oblique to rift directions.

The structural features observed in the seismic profile include large-scale, steeply dipping, planar-to-near-planar extensional faults, with a series of closely spaced faults (e.g., Line 1920; Fig. 10.8a) extending east–west across the entire area. Listric faults (Faults 1a and b, 6, and 10; Fig. 10.8a) are also present, where fault 1 is located at the western and eastern margins, forming a regional graben master known as boundary

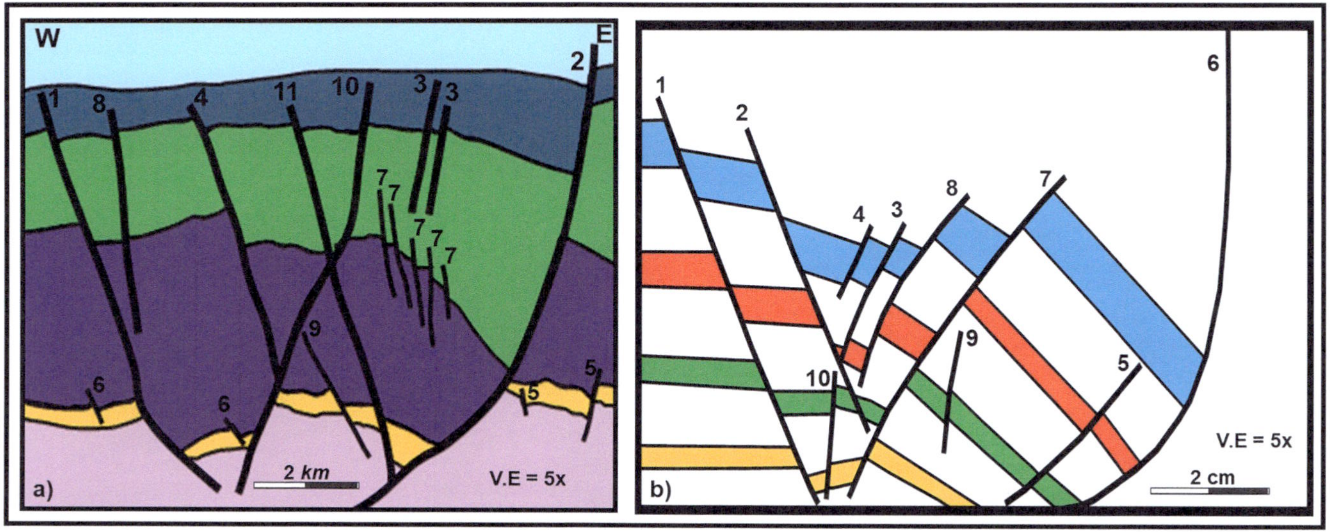

Fig. 10.6 **a** Localised area of interpreted seismic profiles showing different structural styles on X-line 21,920. Faults 1 and 2 represent boundary faults of an apparently planar fault on the left and an apparent listric fault on the right, respectively. Conjugate faults (e.g., X-shaped faults 10 and 11; Y-shaped faults 4 and 10), rotated fault blocks (faults 6 and 7), and apparent planar faults (faults 3, 5, 8, and 9). **b** Analogue model of McClay and Ellis (1987) showing the structural styles developed under extension controlled by a listric fault. Faults 1 and 6 represent boundary faults of planar (left side) and listric (right side), respectively. Y-shaped conjugate faults (faults 2 and 3), rotated fault blocks (faults 7, 8, 3, and 4), and planar faults (faults 9 and 10) are nucleated in the boundary faults of analogue sandbox experiments of McClay and Ellis (1987)

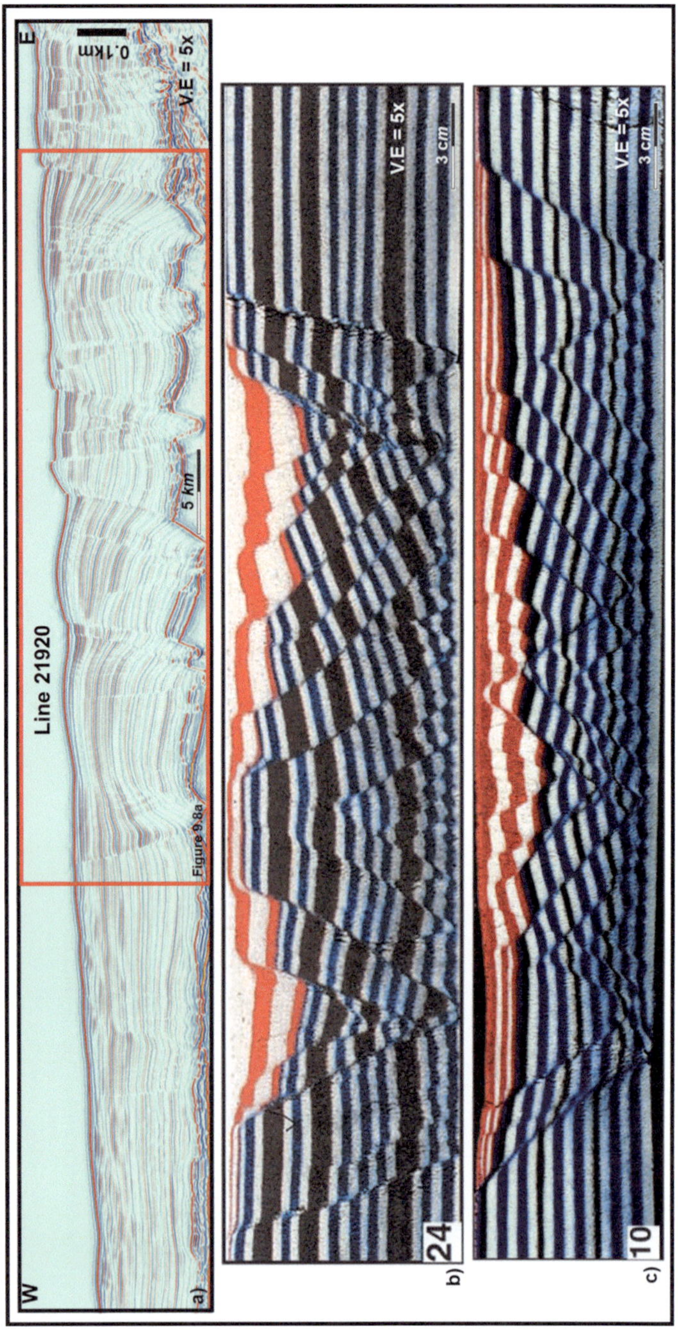

Fig. 10.7 **a** Uninterpreted seismic section of Line 21,920 of the Levantine Margin, **b** Serial section of Line 24 of 60° Oblique Rift (McClay et al., 2002), **c** Serial section of Line 10 of 45° Oblique Rift (McClay et al., 2002)

faults. In the 60º and 45º oblique rift analogue model (McClay et al., 2002), the boundary fault appears as a steeply dipping planar fault (faults 1a and b; Fig. 10.8b, c). No apparent listric faults are observed in the analogue oblique rifts. Boundary faults host nucleated faults with different styles.

Steep, planar faults are the dominant type (Faults 25, 8–17; Fig. 10.8a). They are also found in the analogue model of a 60º oblique rift (faults 1–19; Fig. 10.8b) and in a 45º oblique rift basin (faults 1–18; Fig. 10.8c).

The presence of tilted (rotated) fault blocks correlates with steeply dipping planar faulting (fault 16), which is synthetically oriented and dips east (see Fig. 10.8a). Unlike the McClay et al. (2002) analogue model, which shows both 60º (faults 4, 5, 6, and 7; see Fig. 10.8b) and 45º (fault 12; see Fig. 10.8c), the tilted fault blocks observed are associated with near-planar faulting.

Several conjugate faults (faults 5, 7, and 8; 12 and 14; see Fig. 10.8a) are observed developing within the master graben of the Levantine Margin. However, these conjugate faults are also present in the analogue model by McClay et al., 2002, for both 60º (faults 2 and 13; 7 and 11; 8 and 11; 9 and 11; see Fig. 10.8b) and 45º (faults 5 and 16; 7; 8; 13 and 14; 18 and 1; see Fig. 10.8c) oblique rift basins.

Cross-cutting faults (faults 13 and 14; see Fig. 10.8a) are often associated with conjugate faults (faults 12 and 14; 13 and 1; see Fig. 10.8a). These vertical fault relationships are similar to those in McClay et al., 2002's analogue model for 60º (faults 10 and 11; 10 and 1220; related conjugate faults 8 and 11; 9 and 11; 10 and 1; see Fig. 10.8b) and 45º (faults 9 and 10; 9 and 11; 17 and 11; conjugate faults 9 and 12; 17 and 12; 10 and 1; see Fig. 10.8c) oblique rift basins.

Grabens (faults 1a, 2b; 1a, 3; 1, 6; 1, 8; 2a, b; 10, 14; 10, 1b; Fig. 10.8a) are high- to low-angle, near-planar to listric normal faults mainly associated with horsts (faults 6, 7, 8, 10; Fig. 10.8a) resulting from rifting, indicating extensional tectonics at the Levant Margin caused by localised stresses. Block faults in the analogue model by McClay et al., 2002, at 60º (graben- 2, 3; 5, 12; 6, 12; 7, 12; horst faults 3, 4, 16; Fig. 10.8b) and 45º (graben- 3, 15; 4, 11; 5, 11, 4; 12d, 4; 12e, 5; 12d, 5; 12e, 4; 14, 5, 13; horst faults 15, 4, 13; Fig. 10.8c) are low-angle, near-planar faults (Table 10.1).

The Levantine Margin shares features with McClay et al.'s (2002) oblique-rift models, such as boundary faults, rotated blocks, graben–horst systems, and detachment deformation. It is more complex, with listric faults and asymmetric basins, due to stratigraphic heterogeneity and depth-dependent rheology. These points emphasise the role of sandbox models and seismic data in understanding structural complexity.

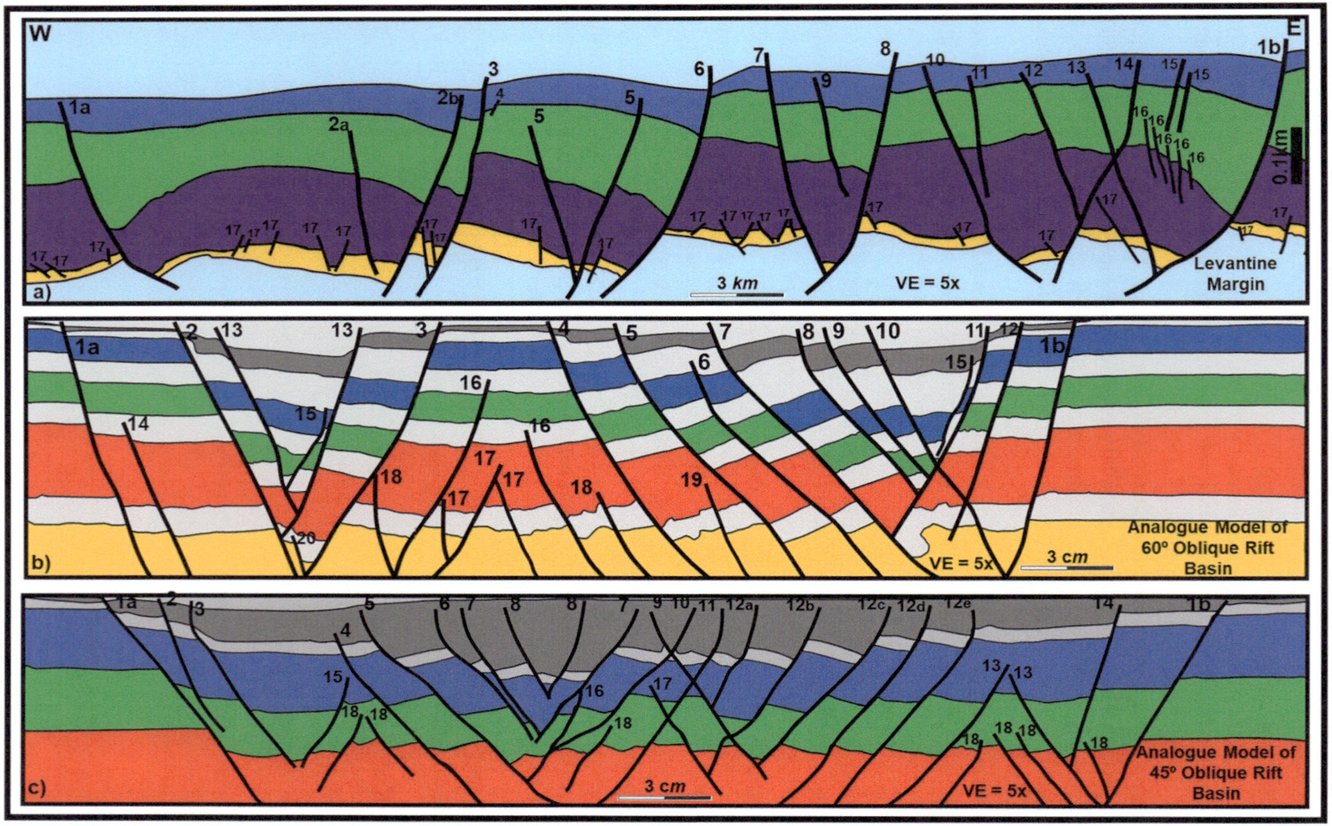

Fig. 10.8 Cross-section of the interpreted segment showing **A** a selected seismic section from X-line 21,920 of the Levantine Margin in 8-1a; **b** analogue model of Line 24 in the 60° Oblique Rift (Modified after McClay et al., 2002); **c** analogue model of Line 10 in the 45° Oblique Rift (Modified after McClay et al., 2002)

Table 10.1 Comparative summary of structural styles on 2D profiles of the Levantine Margin, analogue sandbox experiments by McClay and Ellis (1987) and McClay et al. (2002)

Structural Style	Levantine Margin (lines 22,000, 21,920, 21,400)	McClay & Ellis (1987)	McClay et al. (2002)
Boundary (master) faults	Steeply dipping planar to listric faults; cut Plio-Quaternary and Messinian units; commonly asymmetric	Steep planar boundary faults defining grabens	Steep planar (60°) to near-planar (45°) boundary faults defining oblique rift basins
Planar normal faults	Abundant; steep, evenly spaced; dominant internal graben structure	Dominant fault type: evenly spaced planar faults	Common; steep planar faults present but less dominant than near-planar faults
Near-planar faults	Rare; transitional geometries	Rare	Dominant internal fault style in both 45° and 60° oblique rifts
Listric normal faults	Commonly occur at basin margins, sole into Messinian levels	Developed with progressive extension; associated with rotated layering	Absent
Rotated (tilted) fault blocks	Commonly associated with steep planar and listric faults	Common; progressive rotation with increasing extension	Common; mainly associated with near-planar faults
Graben geometry	Well developed; commonly asymmetric	Well developed; generally symmetric	Well-developed; oblique grabens
Horst blocks	Commonly associated with block faulting	Present but less pronounced	associated with low-angle block faults
Detachment surfaces	Prominent at the Messinian level; brittle–ductile transition	Implicit basal detachment	Explicit basal detachment controlling extension
Conjugate faults (X-, Y-, V-shaped)	Abundant; nucleated within grabens and boundary faults	Present but limited	Well-developed, conjugate arrays
Cross-cutting faults	Common; associated with conjugate systems	Rare	Common in 45° oblique rift profiles
Fault spacing pattern	Variable; influenced by stratigraphy and inheritance	Relatively uniform	Regular; controlled by experimental boundary conditions
Overall profile-scale complexity	High; multiple fault styles coexist on single profiles	Moderate; progressive but simplified fault evolution	Moderate–high; complexity enhanced by oblique extension

10.3 Comparison of Fault Geometry and Interactions of Map View of the Levantine Margin and Analogue Models

This section reviews structural styles in map view, including interaction processes such as relay development, fault linkage, conjugate arrays, and cross-cutting relationships that are difficult to interpret on individual seismic sections. These structural features are compared with the 60° (see Fig. 10.9) and 45° (see Fig. 10.10) oblique-rift models of McClay et al. (2002), providing a reference for understanding fault growth during oblique extension.

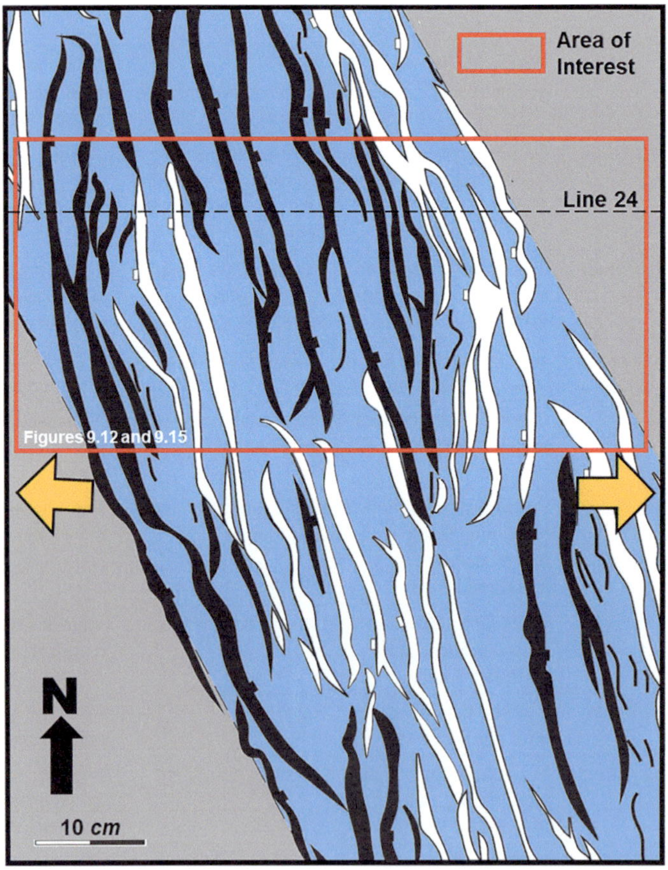

Fig. 10.9 Map view of the analogue model of a 60° oblique rift, showing the interpretation of the fault pattern in a line diagram at the extension's end. Light-band faults dip to the left, while dark-band faults dip to the right. The blue shading indicates the stretched rubber sheet at the base of the model. The red rectangle marks the area of interest (Figs. 10.11, 10.12) (after McClay et al., 2002)

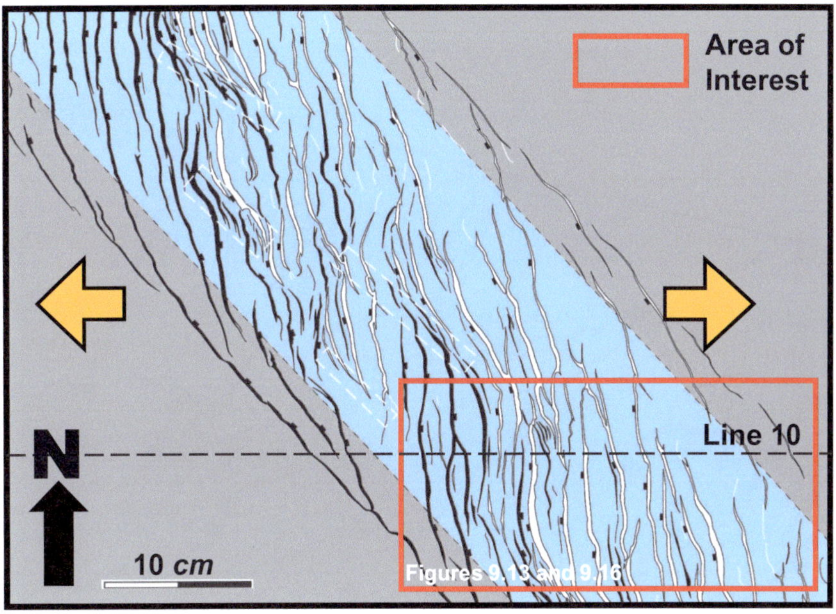

Fig. 10.10 Map view of a 45° oblique rift's analogue model, showing fault patterns in a line diagram. Light-band faults dip left, dark-band faults dip right. Blue shading indicates stretched rubber at the base. The red rectangle marks the selected area (McClay et al., 2002)

10.3.1 Fault Length, Shape, and Tip Geometry

Fault lengths range from short to long, with mega-faults displaying displacement in a specific direction. Long faults show significant displacement gradients, while shorter faults are densely distributed and accommodate local strain. The predominant trend is landward, as meso-faults are evenly distributed between basinward and landward directions (see Fig. 10.11). These faults mainly strike obliquely in the NNE-SSW direction. However, in the 60° (see Fig. 10.12) and 45° (see Fig. 10.13) Oblique Rift Models, fault lengths include mesoscopic and megascopic faults that predominantly strike obliquely in the NNW-SSE direction (e.g., McClay et al., 2002).

The fault tip lines are elliptical or lenticular and are oriented at steep angles relative to the extension direction (see Fig. 10.11). These shapes suggest progressive fault growth rather than sudden connection. Such fault tips resemble those in the 60° (see Fig. 10.12) and 45° (see Fig. 10.13) Oblique Rift Models (e.g., McClay et al., 2002). The similarity between natural and experimental fault tips supports the idea that displacement-driven growth is a key process.

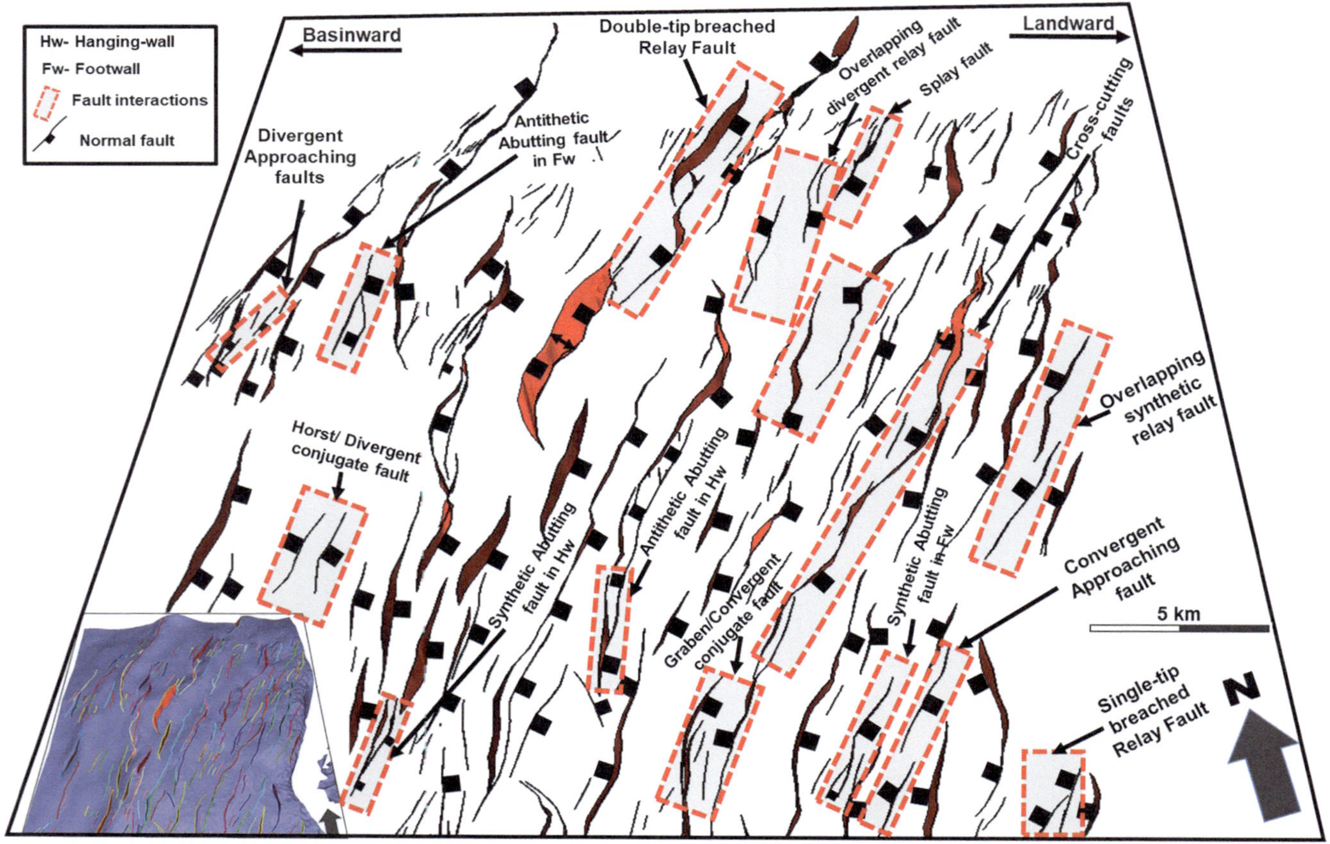

Fig. 10.11 Fault Map of FM-3 of the Levantine Margin, showing the geometric relationships between interacting faults highlighted in red, broken, shaded rectangles

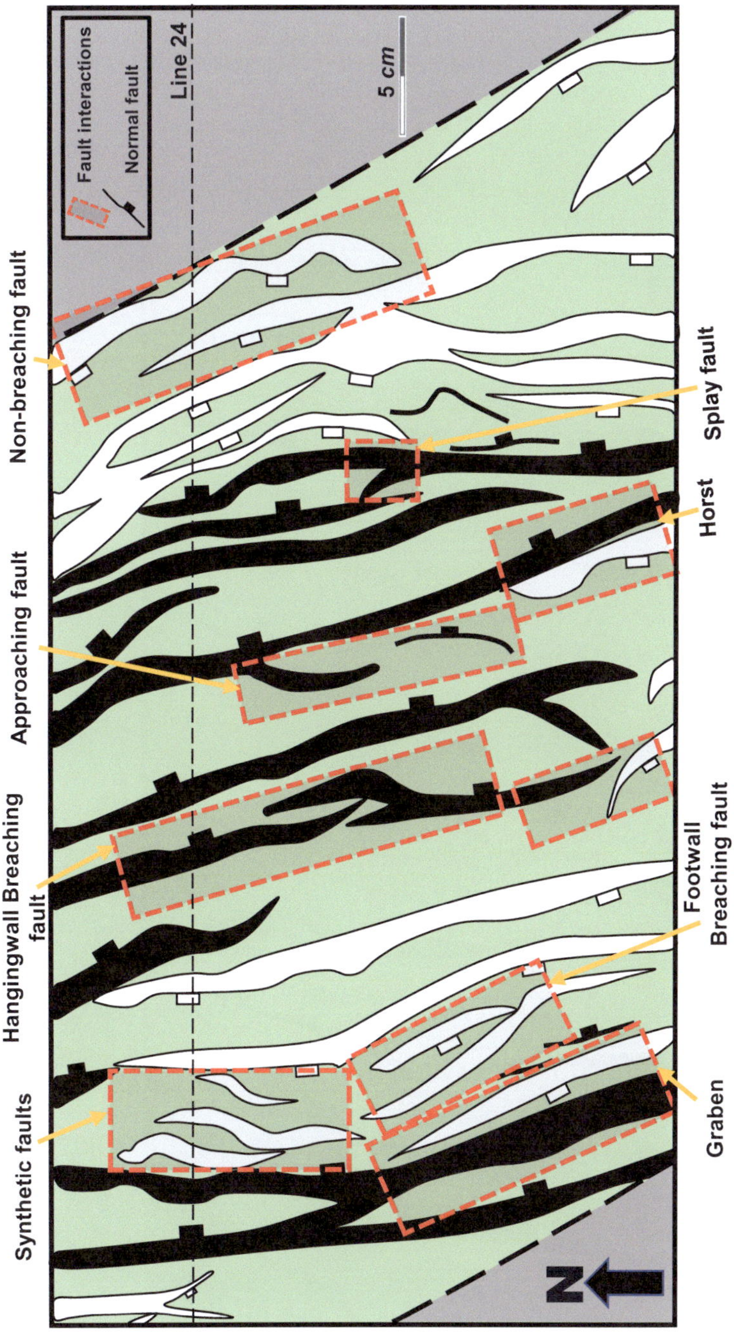

Fig. 10.12 Analogue Model of a 60° Oblique Rift showing the geometric relationships between interacting faults, highlighted in red broken-shaded rectangles. The white-shaded faults represent the synthetic faults, while the black-shaded faults denote the antithetic faults (modified after Mc Clay et al. 2002)

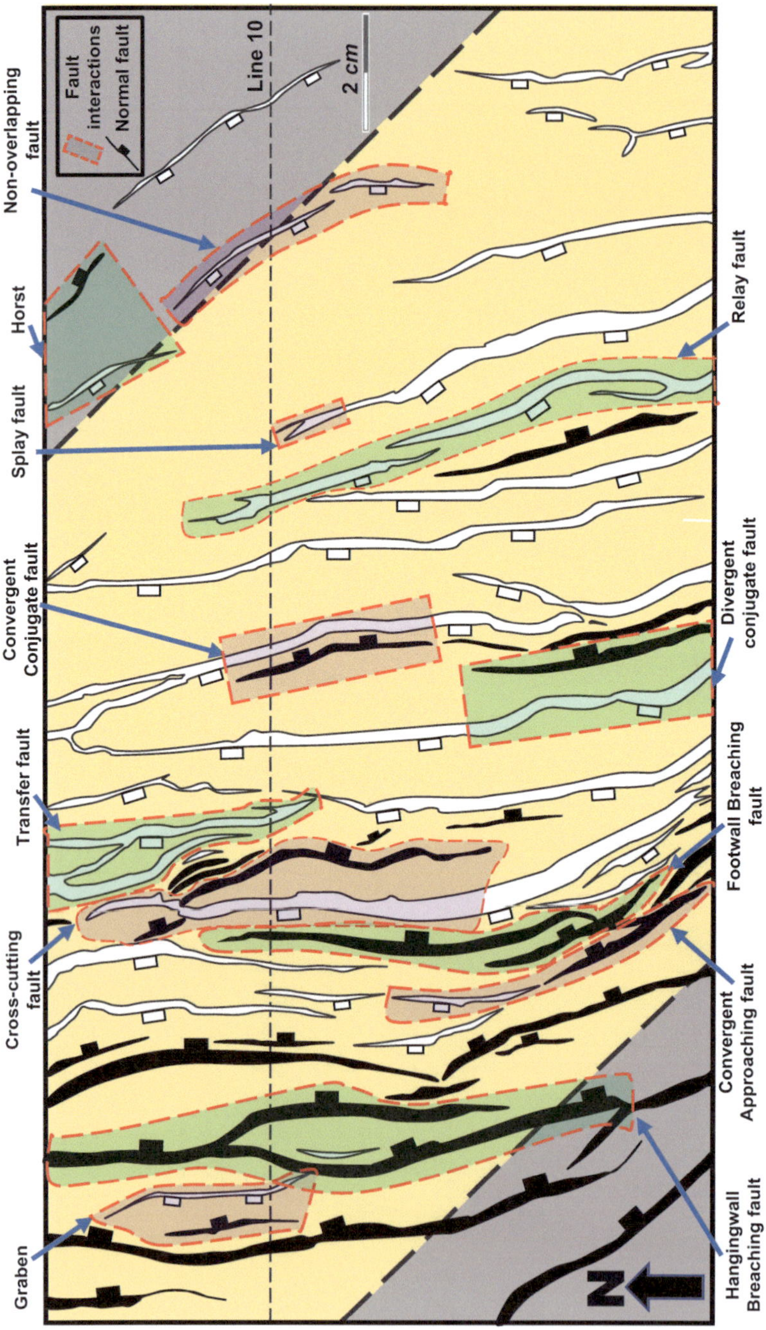

Fig. 10.13 Analogue Model of a 45° Oblique Rift illustrating the geometric relationships between interacting faults, highlighted in red broken-shaded rectangles. The white-shaded faults represent the synthetic faults, while the black-shaded faults denote the antithetic faults (modified after Mc Clay et al. 2002)

10.3.2 Conjugate Fault Systems and Map-Scale Organisation

Conjugate fault systems were observed with faults dipping in opposite directions, often converging or diverging locally and linked to graben–horst structures (see Fig. 10.11). The 60° and 45° (see Figs. 10.12 and 10.13) oblique-rift models feature interlocking intra-rift conjugate faults perpendicular to extension, with more variation in their orientation and spacing than in natural systems, reflecting heterogeneous stress and stratigraphic factors absent in sandbox experiments. These conjugate systems are mostly convergent, though occasionally divergent.

Seismic fault maps in the Levantine Margin show complex, multiscale faults with segmentation, elliptical tips, relay development, conjugate arrays, and extension-driven strain. Faults vary in polarity, interaction, and linkage due to stratigraphy, inherited structures, and rheology. While sandbox models aid interpretation, 3D seismic data is essential for full understanding.

10.3.3 Geometrical Relationships of Interacting Faults

Fault interaction styles include relay, approaching, abutting, splay, and cross-cutting faults (see Fig. 10.11). These fault networks display a wider variety of relay geometries. Well-developed relay zones range from non-breached overlaps to fully breached relays, with single- and double-tip geometries. These structures often show synthetic and antithetic configurations relative to bounding faults. In analogue oblique-rift models, relay faults are common but tend to be more uniform, with overlaps mainly being synthetic and breaching through single-tip linkage, due to simplified boundary conditions (see Figs. 10.12 and 10.13). Approaching faults can converge or diverge, and their interactions with smaller subsidiary faults increase with greater displacement (see Fig. 10.11). Most models exhibit convergent faulting behaviour, consistent with experimental findings (see Figs. 10.12 and 10.13) (Table 10.2).

10.3.4 Kinematic Relationships of Interacting Faults

Kinematic relationships of interacting faults on the Levantine Margin, and McClay et al. (2002) oblique-rift experiments, emphasise displacement sense, fault trace orientation, and bisector geometry, thereby constraining extension direction and magnitude during fault growth and linkage. Kinematic analysis of the Levantine Margin uncovers varied interaction patterns influenced by fault-plane orientation and

Table 10.2 Comparative summary of structural styles on map view of the Levantine Margin and analogue sandbox experiments by McClay et al. (2002)

Structural style (map view)	Levantine margin (FM-3)	Analogue sandbox model—McClay et al. (2002)
Overall fault pattern	Dense, heterogeneous fault network; multiple generations of interacting faults	Organised fault arrays controlled by imposed oblique extension
Fault length distribution	Broad spectrum from short, isolated segments to regional-scale mega-faults	Mesoscopic to megascopic faults dominate; limited short segments
Fault shape	Curvilinear to segmented traces; irregular geometries are common	Relatively smooth, continuous fault traces
Fault tip geometry	Elliptical and lenticular tips; abrupt and tapered terminations	Elliptical to lenticular tips; gradual tapering
Fault strike orientation	Predominantly NNE–SSW; oblique to extension direction	Predominantly NNW–SSE; oblique to extension direction
Boundary/master faults (map view)	Long, continuous master faults defining graben margins	Long, continuous master faults define oblique rift basins
Internal fault segmentation	Strong segmentation, overlapping and en-échelon arrangements are common	Moderate segmentation; more uniform spacing
Conjugate fault systems	Both convergent and divergent conjugate arrays present	Interlocking intra-rift conjugate systems; predominantly convergent
Relay zones	Widespread; synthetic and antithetic relays; non-breached and breached (single- and double-tip)	Common; dominantly synthetic relays; mainly non-breached or single-tip breached
Approaching fault patterns	Convergent and divergent approaches observed	Convergent approaching faults dominate
Abutting faults	Present; faults terminate against larger structures	Absent
Splay faults	Common; branching from master or relay faults	Present, especially in 60° oblique rift models
Cross-cutting faults	Common; diagonal propagation and overprinting relationships	Common, particularly in 45° oblique rift models
Graben–horst expression (map view)	Irregular graben–horst patterns; asymmetric blocks	Regular graben–horst patterns; relatively symmetric blocks

(continued)

Table 10.2 (continued)

Structural style (map view)	Levantine margin (FM-3)	Analogue sandbox model—McClay et al. (2002)
Fault interaction density	High; multiple interaction styles coexist within small areas	Moderate; interaction styles are more uniform
Overall map-scale complexity	Very high; reflects stratigraphic heterogeneity and structural inheritance	Moderate–high; complexity driven mainly by oblique extension geometry

displacement vectors. Intersections create antithetic, synthetic, and complex interactions (see Fig. 10.14). The orientation of the acute bisectors consistently indicates extension-dominated strain regimes. In analogue oblique-rift models, interacting faults are mainly synthetic–synthetic, with intersection lines parallel or oblique to the displacement direction (see Figs. 10.15 and 10.16). In natural and experimental systems, acute bisectors align with extension margins, indicating shared kinematic control despite differences in fault polarity.

i. Displacement Sense and Polarity Pairing

Fault intersections show multiple polarity pairs (see Fig. 10.14), classified kinematically as:

- Synthetic–Synthetic (S–S): faults dip in the same direction, with compatible displacement vectors.
- Antithetic–Synthetic (A–S): faults dip in opposite directions, and displacement vectors are nearly perpendicular to the intersection line.
- Antithetic–Antithetic (A–A): faults dip away from each other, creating opposing displacement fields and localised strain.

Quantitatively, S–S interactions dominate in mature relay zones and basin-bounding faults with the largest displacements. A–S and A–A interactions are typical in fault nucleation zones and graben interiors, where strain is distributed, and displacements are lower. This pattern shows that faults become progressively simpler in kinematics as they develop and connect.

ii. Intersection Geometry and Acute Bisector Analysis

The bisector between fault traces shows the main strain field. In the Levantine Margin, bisector directions align with regional extension, confirming that deformation is extension-driven.

- When the acute bisector aligns with the extension, interactions are typically dominated by synthetic processes, improving displacement transfer and fault linkage.
- When the bisector is oblique or rotated, mixed-polarity interactions predominate, indicating stress variations caused by nearby faults or mechanical layering.

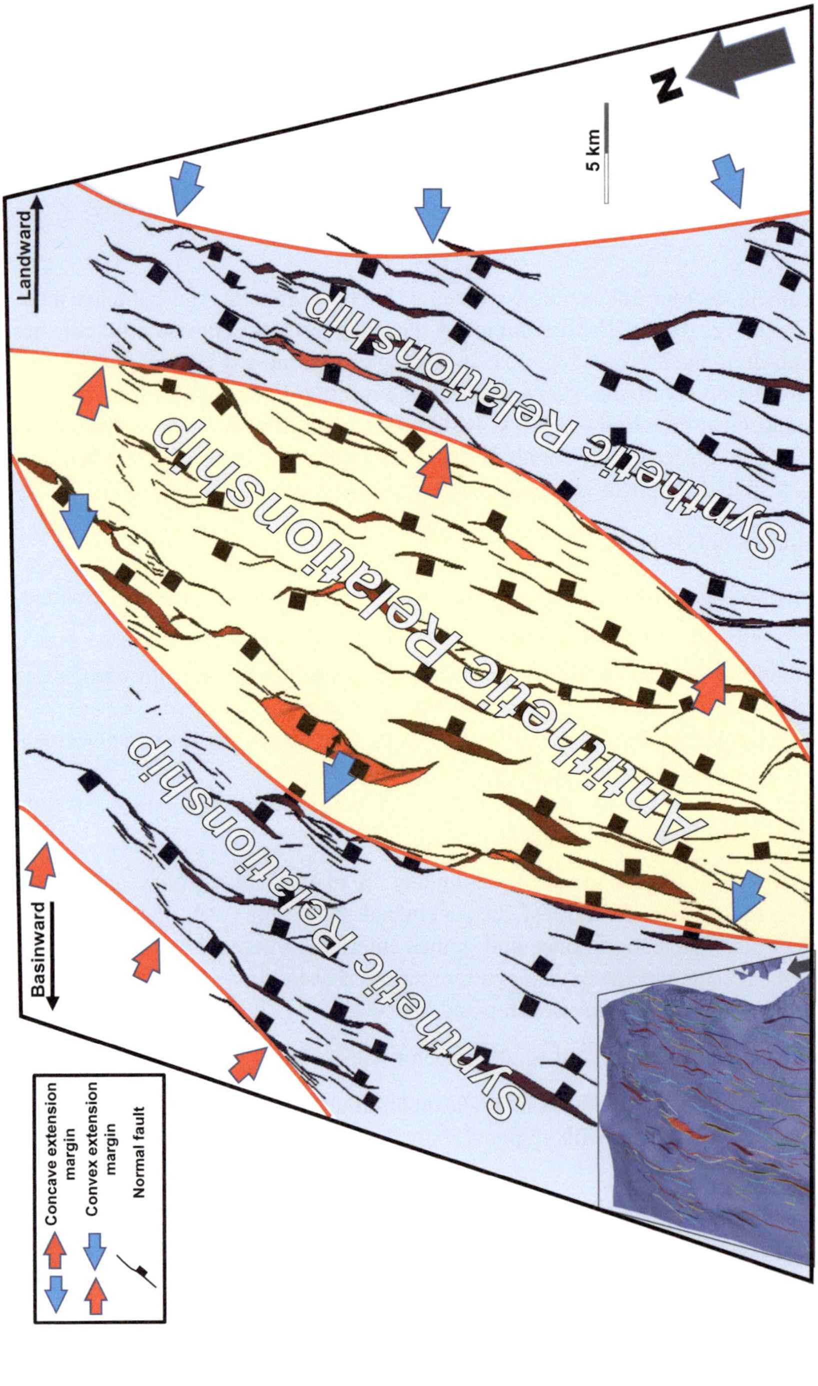

Fig. 10.14 Fault Map of FM-3 in the Levantine Margin, showing the Antithetic and Synthetic relationships of dominant acute bisectors between interacting faults. The green arrows indicate dominant extension margins

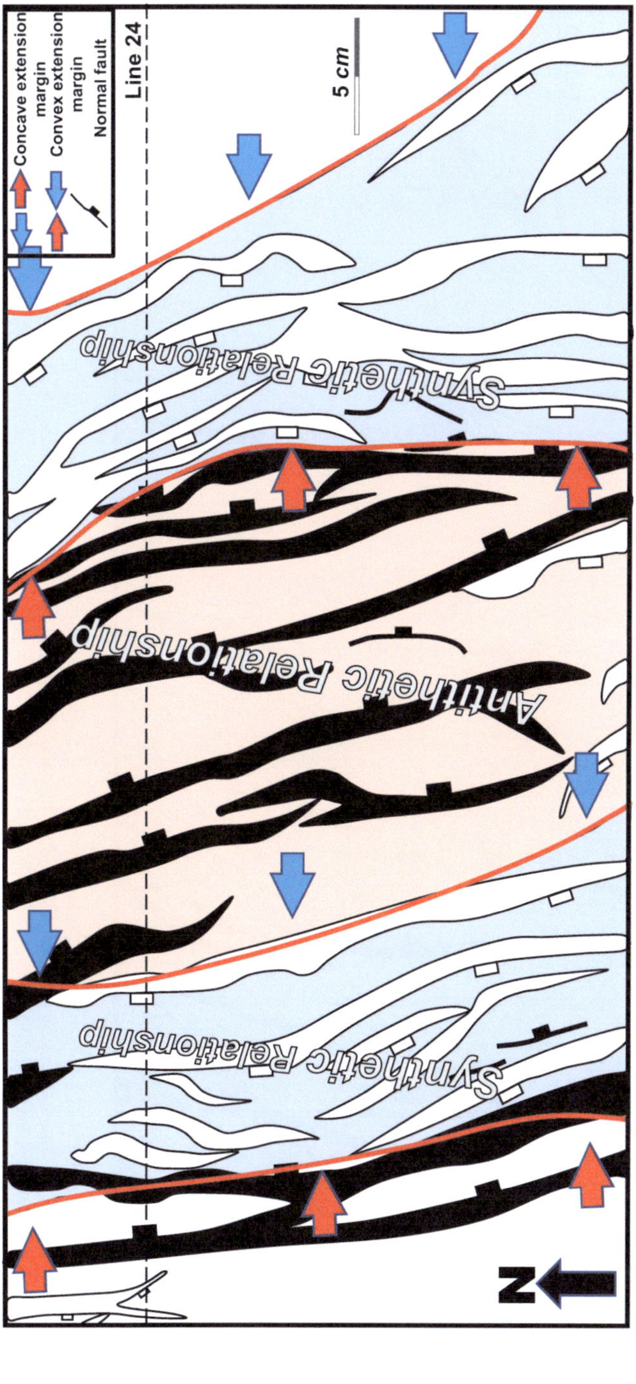

Fig. 10.15 Analogue model of a 60° oblique rift illustrating the antithetic and synthetic relationships of dominant acute bisectors between interacting faults. The red arrows indicate the main extension margins. The white-shaded faults represent the synthetic faults, while the black-shaded faults denote the antithetic faults (modified after Mc Clay et al. 2002)

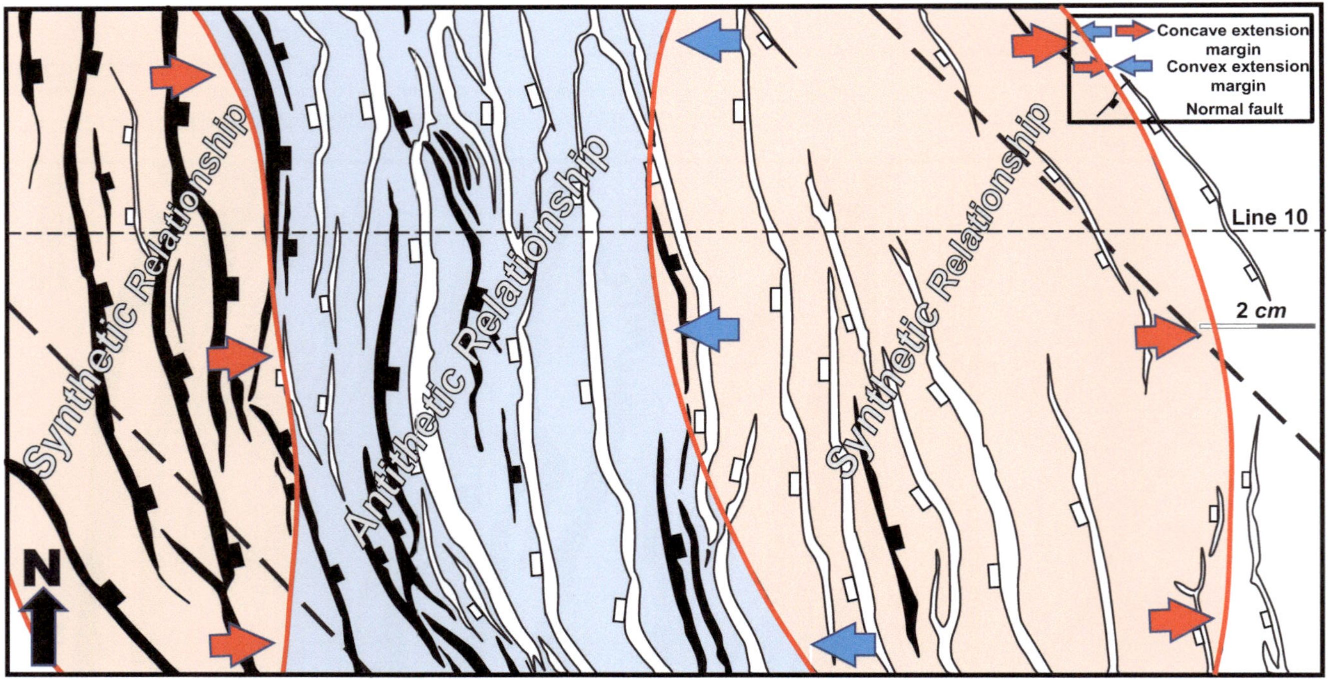

Fig. 10.16 Analogue Model of a 45° Oblique Rift illustrating the Antithetic and Synthetic relationships of dominant acute bisectors between interacting faults. The red arrows show the main extension margins. The faults shaded in white represent the synthetic faults, while those shaded in black denote the Antithetic faults (Modified after Mc Clay et al. 2002)

In analogue oblique-rift models, acute bisectors align more consistently with the boundary conditions (see Figs. 10.15 and 10.16; Table 10.3).

McClay et al. (2002) *models show synthetic interactions on faults crossing relay ramps during oblique extension. Acute bisectors align with extension, and displacement vectors change little. The Levantine Margin shows a wider range with more A–S and A–A intersections early in fault development and layered intervals, due to persistent antithetic components and complex geometries.*

10.4 Conclusion

This chapter stresses the importance of combining 3D seismic interpretation with analogue sandbox modelling to understand how extensional fault systems develop at continental margins. Comparing the Levantine Margin with analogue experiments shows that many fault features observed in seismic data, such as planar and listric faults, rotated blocks, graben–horst structures, detachment surfaces, and conjugate faults, are linked to mechanically driven processes.

It highlights differences between natural and analogue systems. The Levantine Margin displays various fault polarities and interaction types, including mixed relationships, complex relay geometries, and diverse linkage pathways. These features, shaped by stratigraphy, inherited structures, and rheology, create kinematic complexities that exceed what simplified models can represent. Quantitative analysis indicates that deformation is mainly extensional in both systems, but natural systems have a longer, more complex history.

Learning Objectives

After completing this chapter, readers will be able to:

1. Compare fault architectures of the Levantine Margin with analogue sandbox model experiments.
2. Identify and classify fault geometries, including planar and listric faults, rotated blocks, graben–horst systems, detachments, and conjugate arrays.
3. Explain how faults nucleate, propagate, interact, and link through time using integrated seismic and analogue observations.
4. Distinguish between non-breached, single-tip breached, and complex relay ramps in natural and experimental fault systems.
5. Interpret acute bisector geometry and intersection orientations to infer extension direction and displacement transfer efficiency.
6. Critically assess the strengths and limitations of analogue sandbox models as mechanical representations of natural fault networks.
7. Apply comparative seismic–analogue insights to improve fault interaction and kinematic analysis in other rifted continental margins.

Table 10.3 Comparative kinematic characteristics of fault interactions derived from 3D seismic fault maps of the Levantine Margin and analogue sandbox experiments by McClay et al. (2002). The comparison emphasises displacement sense, polarity pairing, relay evolution, and acute bisector geometry under extensional strain

Kinematic attribute	The Levantine Margin (Natural System)	Analogue sandbox models (McClay et al., 2002)	Interpretative significance
Dominant fault polarity pairing	Mixed Synthetic–Synthetic (S–S), Antithetic–Synthetic (A–S), and Antithetic–Antithetic (A–A) interactions	Predominantly Synthetic–Synthetic (S–S)	Natural systems preserve early-stage polarity diversity; analogue models capture mature, strain-localised end members
Spatial distribution of interaction types	A–S and A–A common within graben interiors and nucleation zones; S–S dominant along basin-bounding faults	S–S interactions are dominant throughout the rift domain	Indicates progressive kinematic simplification with fault growth in natural systems
Intersection line orientation	Parallel, oblique, and locally perpendicular to the displacement direction	Mainly parallel or oblique to the displacement direction	Greater orientation variability reflects heterogeneous stress fields in the natural system
Acute bisector orientation	Generally aligned with extension but locally dispersed	Consistently aligned with extension	Dispersion in natural systems records stress perturbations from neighbouring faults and layering
Relay ramp kinematics	Non-breached, single-tip breached, and complex multi-segment relays; mixed polarity during early stages	Mostly single-tip breached relays with S–S kinematics	Natural relays evolve through multiple kinematic stages before linkage
Fault linkage style	Progressive linkage with temporary displacement partitioning across segments	Rapid linkage with efficient displacement transfer	Highlights the role of time-dependent growth in natural fault networks
Displacement continuity along strike	Increases markedly after relay breaching; variable before linkage	High continuity once linkage is initiated	Confirms threshold-controlled transition from segmented to through-going faults

(continued)

Table 10.3 (continued)

Kinematic attribute	The Levantine Margin (Natural System)	Analogue sandbox models (McClay et al., 2002)	Interpretative significance
Role of mechanical stratigraphy	Strong influence; promotes antithetic components and kinematic complexity	Implicit and uniform	Stratigraphic heterogeneity enhances kinematic diversity at the seismic scale
Overall kinematic regime	Extension-dominated with locally complex strain partitioning	Extension-dominated with simplified strain geometry	Both systems are governed by extension, but natural systems retain higher-order complexity

References

McClay, K. R., & Ellis, P. G. (1987). Geometries of extensional fault systems developed in model experiments. *Geology, 15*(4), 341–344.

McClay, K. R., Dooley, T., Whitehouse, P., & Mills, M. (2002). 4-D evolution of rift systems: Insights from analogue modelling. *AAPG Bulletin, 86*(6), 935–959.

Chapter 11
Synthesis

Abstract This chapter synthesises the methodological, structural, quantitative, and comparative understandings developed throughout the textbook to determine how a 3D seismic interpretation workflow improves the reliability, reproducibility, and geological application of subsurface models. Using the Levantine Margin as a type locality, it integrates structural–stratigraphic frameworks, fault identification criteria, quantitative fault scaling, analogue model comparisons, and regional margin analysis to address both objective and subjective uncertainties inherent in seismic interpretation. The chapter shows that horizon-first workflows, resolution-aware fault mapping, and multi-scale validation reduce interpretive bias and improve structural coherence. Quantitative analyses of scaling relationships and displacement gradients reveal differences in fault maturity, interaction intensity, and mechanical stratigraphy across structural domains. Comparative evaluation with other rifted margins demonstrates that mechanical layering exert stronger control on rift architecture than extension magnitude alone. By bridging analogue experiments, seismic-scale observations, and uncertainty-aware practices, this synthesis advances a reproducible, process-based framework for interpreting extensional fault systems. Emphasis is placed on how advanced seismic interpretation supports energy-transition applications, including carbon capture, utilisation and storage (CCUS), geothermal systems, and subsurface risk reduction. The chapter concludes by revealing future directions for machine learning and subsurface modelling.

Keywords Seismic uncertainty · Modelling · Quantitative analysis · Geothermal · CCUS

K. Chiadikobi, *Advances in Horizon Mapping and Fault Interpretation*,
https://doi.org/10.1007/978-3-032-23917-4_11

11.1 Introduction

This chapter demonstrates that a seismic interpretation workflow significantly improves the reliability, reproducibility, and geological realism of fault and stratigraphic models derived from 3D seismic data. By integrating qualitative interpretation, structural–stratigraphic frameworks, quantitative fault analysis, and comparisons with natural and analogue sandbox models, the Levantine Margin addresses both the sources of uncertainty and their effects in seismic interpretation. The discussion summarises Chaps. 1–10 and places them within the broader context of structural geology, seismic interpretation methods, and subsurface energy applications.

11.2 Uncertainties in Seismic Interpretation

Uncertainty is inherent to seismic interpretation, stemming from both the limitations of seismic data and the interpreter's cognitive processes. This chapter shows that uncertainty does not simply reflect poor interpretive practice; rather, it results from the interaction of data quality, geological complexity, and interpretive decision-making.

Seismic interpretation uncertainty are broadly divided into objective and subjective components (Tannert et al., 2007; Bond, 2015; Alcalde & Bond, 2022; Andrews et al., 2024). Objective uncertainty originates from the seismic data and is influenced by acquisition parameters, processing workflows, velocity model accuracy, and the fundamental limits of seismic resolution (Faleide et al., 2021; Michie et al., 2021; Pérez-Díaz et al., 2020; Steventon et al., 2022). These factors limit the interpreter's ability to accurately image subsurface structures, especially in regions characterised by complex fault geometries, steep dips, or mechanically heterogeneous stratigraphy.

Objective uncertainty was most notable in deeper stratigraphic layers and in zones affected by chaotic or low-coherence reflections. Resolution limitations hindered the detection of small-offset faults, aligning with seismic modelling studies that show fault displacements below the tuning thickness are unlikely to be reliably imaged (Dimmen et al., 2023; Faleide et al., 2021). Errors in velocity modelling further increase uncertainty during time-to-depth conversion, where even modest velocity deviations can lead to significant depth errors, especially at greater two-way travel times (Michie et al., 2021; Pérez-Díaz et al., 2020).

Subjective uncertainty stems from the interpreter and reflects personal experience, conceptual frameworks, and interpretive bias (Tannert et al., 2007; Alcalde & Bond, 2022). This type of uncertainty is most evident when mapping discontinuous reflectors or subtle fault offsets, where multiple geological explanations are equally plausible. In such situations, the interpreter must make judgment calls regarding reflector continuity, fault linkage, and structural significance. Previous research has demonstrated that different interpreters can produce significantly different structural models from the same seismic dataset, even when applying similar technical workflows (Alcalde & Bond, 2022; Bond et al., 2007).

This textbook illustrates that workflow design plays a vital role in limiting subjective uncertainty. By adopting a horizon-first interpretation strategy, the workflow curtailed interpretive freedom during fault mapping. Stratigraphic surfaces offered continuous reference frameworks against which fault offsets could be tested and validated across multiple horizons. Fault interpretations that could not be aligned with consistent stratigraphic displacement were amended or rejected, thereby reducing over-interpretation.

The absence of well control adds to uncertainty. Without velocity calibration derived from wells, depth conversion depends on regional velocity models, which increases uncertainty in absolute depth estimates. This challenge is common in frontier or deep-water basins and highlights the importance of transparent reporting of assumptions and uncertainty limits in structural models (Michie et al., 2021; Pérez-Díaz et al., 2020).

Recent advances promote moving beyond deterministic interpretations towards explicit uncertainty quantification, including sensitivity testing, probabilistic depth conversion, and ensemble-based interpretation methods (Andrews et al., 2024; Bond, 2015; Steventon et al., 2022). Although these methods are beyond the scope, the results clearly show that even deterministic workflows can significantly reduce uncertainty when carefully designed and rigorously applied.

11.3 Structural–Stratigraphic Frameworks

Structural–stratigraphic frameworks provide the foundation for reliable seismic interpretation by integrating stratigraphy and structure into a single, internally consistent model. This chapter demonstrates that interpretation accuracy improves significantly when stratigraphic surfaces are established before fault delineation and when regional and local datasets are interpreted together within a unified framework.

A horizon-mapping workflow limits fault interpretation by applying stratigraphic continuity and displacement rules that faults must meet to be considered geologically plausible. By mapping laterally continuous horizons before fault tracing, the interpreter creates reference surfaces against which reflector offsets, fault throw consistency, and fault continuity can be assessed. This method reduces the risk of misinterpreting noise, processing artefacts, or stratigraphic terminations as faults, a challenge widely acknowledged in seismic interpretation practice (Bacon et al., 2003; Brown, 2012; Onajite, 2014).

The effectiveness of this approach is particularly evident in areas with complex structures where reflector continuity is locally interrupted. In such regions, horizons help to confirm fault interpretations across multiple stratigraphic levels. Faults that do not exhibit consistent offsets on neighbouring horizons can be dismissed or refined, improving the robustness of the final fault network. Earlier studies have shown that interpretation reliability increases when fault geometries are evaluated in relation to stratigraphic architecture rather than interpreted in isolation (Brown, 2012; Carvajal et al., 2023).

Integrating regional 2D seismic profiles with detailed 3D seismic interpretation further improves structural–stratigraphic understanding. Regional 2D data capture large-scale basin architecture and long-wavelength structural trends that may not be immediately obvious in 3D datasets with limited spatial coverage. When these regional insights are incorporated into 3D interpretation, they provide essential boundary conditions that guide fault correlation, segmentation analysis, and structural domain definition (Bacon et al., 2003; Onajite, 2014).

When applied carefully, vertical exaggeration also strengthens structural–stratigraphic frameworks. Moderate exaggeration highlights subtle dip changes, monoclines, and small offsets, all of which are vital for recognising early fault development and relay structures. However, the Levantine Magin demonstrates that exaggeration must be controlled and cross-checked against true-scale views to prevent geometric distortion, as emphasised in previous work on seismic visualisation practices (Stewart, 2011, 2012; Stone, 1991).

The analogue-to-digital workflow further illustrates the pedagogical and practical value of structural–stratigraphic frameworks. Manual interpretation of selected 2D profiles helped develop early conceptual understanding of stratigraphic stacking patterns and fault relationships, while digital 3D interpretation allowed for precise geometric refinement and validation. This multi-scale, multi-method approach aligns with recommendations from reproducibility studies, which emphasise transparency, cross-validation, and iterative refinement as essential elements of robust subsurface interpretation (Steventon et al., 2022).

11.4 Fault Identification and Structural Styles

Fault identification is one of the most interpretationally sensitive steps in seismic analysis because faults rarely appear as discrete, continuous reflections. Instead, interpreters infer faults from indirect seismic expressions, such as reflector offsets, terminations, dip changes, and amplitude variations. This chapter demonstrates that fault identification accuracy depends on both seismic data quality and the interpreter's ability to apply consistent geological criteria.

Seismic resolution fundamentally restricts fault detectability. Faults with throws below the seismic data's vertical resolution cannot be reliably imaged and may remain undetected or be mistaken for stratigraphic noise (Dimmen et al., 2023; Faleide et al., 2021; Widess, 1973). In this dataset, small-offset faults were most challenging to identify in deeper stratigraphic layers and in zones characterised by chaotic or low-coherence reflections. These limitations highlight that the absence of a fault in seismic data does not necessarily mean it is absent in the geology.

Interpreter criteria greatly affect how faults are recognised and traced. Faults were identified through a combination of reflector-offset consistency, sudden changes in dip, and vertically aligned zones of reflector disruption. Using these criteria helped reduce interpretative uncertainty and prevented the spread of false fault segments. Previous research shows that inconsistent fault-picking criteria are a major source of variability between interpreters working on the same dataset (Bond, 2015; Alcalde & Bond, 2022).

The structural styles identified as planar faults, listric faults, block faulting (grabens, horsts, and tilted blocks), and conjugate fault systems are characteristic of extensional tectonic regimes and reflect variations in mechanical stratigraphy and deformation style (Cartwright et al., 1996; Gibbs, 1984; Lowell, 1985; Shelton, 1984). Planar faults dominate where deformation occurs within relatively homogeneous brittle layers, whereas listric faults develop where displacement localises on mechanically weak horizons, producing curvature and flattening with depth.

Block faulting patterns record variations in extension across the basin. Grabens and horsts form where opposite faults interact symmetrically, while tilted blocks develop when rotation occurs along a dominant fault surface. These configurations reveal spatial differences in strain magnitude and fault spacing, aligning with traditional extensional basin models (Gibbs, 1984).

Conjugate fault systems offer strong evidence for fault interaction and kinematic evolution. V-, Y-, and X-shaped fault geometries indicate different forms of fault linkage, termination, and cross-cutting, and record ongoing deformation rather than single-event faulting (Baudon & Cartwright, 2008; Dawers et al., 1993; Needham et al., 1996). Visualising these relationships in three dimensions reveals interaction zones, relay structures, and segmentation patterns that are challenging to resolve using 2D data alone.

Importantly, this chapter shows that structural styles are not fixed categories but signs of fault-system maturity. Segmented planar faults often develop into linked listric or composite structures as displacement builds up and faults interact. Recognising these evolutionary connections is crucial for understanding basin formation, fault-growth histories, and strain partitioning.

11.5 Three-Dimensional Evolution of Extensional Fault Networks

Extensional fault systems do not develop as isolated, flat structures. Instead, they form three-dimensional, mechanically interacting networks where fault growth, linkage, and strain localisation are driven by the combined effects of geometry, kinematics, and displacement transfer. The Levantine Margin offers a clear seismic-scale example of this behaviour, with faults interacting across multiple stratigraphic levels to create coherent, evolving fault populations rather than separate structures.

In the initial stages of deformation, faults nucleate as segmented, sub-parallel structures that mainly grow through tip propagation. As fault density rises, proximity encourages interaction. These interactions are not random: they occur within transfer zones, where displacement gradients, fault overlap, and stress perturbations concentrate strain. Relay zones develop as the primary mechanism for accommodating displacement between neighbouring faults, acting as temporary buffers that control connectivity during fault growth.

With continued extension, relay zones develop from unbreached overlaps, through single-tip breaching, to double-tip breaching, illustrating a clear trajectory from distributed strain to localised fault linkage. This progression indicates increasing mechanical coupling between fault segments and signifies a transition from fault populations dominated by interaction to systems dominated by through-going connectivity. In three dimensions, this evolution is shown by vertically persistent branch lines, curved fault traces, and narrowing fold widths with depth, reflecting progressive strain localisation within the faulted volume.

Beyond relay systems, additional geometric interaction styles such as approaching, abutting, splay, and cross-cutting faults record distinct mechanical responses to stress redistribution around propagating fault tips. Approaching faults capture arrested growth and stress interactions without linkage, preserving evidence of transient strain accommodation. Abutting and splay faults define fault hierarchy and reveal how displacement is partitioned within branching systems. Cross-cutting faults record late-stage overprinting and provide robust constraints on relative fault timing and network reorganisation.

Kinematic analysis reveals that fault–fault intersection geometry relative to the displacement direction primarily influences linkage efficiency. When intersection lines are parallel to displacement, slip transfer occurs efficiently with minimal wall-rock deformation. Conversely, displacement-perpendicular and mixed kinematic relationships require more wall-rock strain, resulting in complex breaching geometries and curved intersection lines. These configurations are progressively less efficient but are mechanically necessary solutions for accommodating strain as the fault network evolves.

Displacement sense further influences strain distribution within the network. Synthetic interactions, characterised by compatible slip directions, promote efficient linkage and minimal wall-rock strain, while antithetic interactions create kinematic incompatibilities that need extra deformation to accommodate space loss or overlap. The association of convex displacement patterns with synthetic interactions and concave patterns with antithetic interactions offers a predictive framework for recognising strain localisation in map view.

These observations show that extensional fault systems grow through structured, interaction-driven processes rather than randomly. Fault geometry determines interaction points, kinematics govern slip transfer, and displacement–strain relationships influence linkage efficiency. As a result, fault networks mirror cumulative interactions within a stable regional stress field.

11.6 Quantitative Fault Characterisation

Quantitative fault analysis offers an objective framework for assessing fault growth, maturity, and mechanical behaviour, moving beyond purely descriptive interpretation. By measuring fault length, displacement, throw, and displacement gradients, it shows how quantitative metrics complement structural–stratigraphic interpretation and reveal variations within fault populations.

Fault length–displacement relationships are among the most commonly used tools for evaluating fault growth. Numerous studies indicate that normal faults often follow power-law scaling, with maximum displacement predictably increasing with fault length (Dawers et al., 1993; Fossen & Rørnes, 1996; Marrett & Allmendinger, 1991, 1992; Needham et al., 1996). In mechanically simple systems, these relationships suggest self-similar fault growth and a relatively uniform strain distribution. Faults in different structural domains display significantly different scaling behaviour, reflecting variations in fault maturity, interaction, and geological complexity.

Fault domains FM-1 and FM-2 exhibit strong correlations between fault length and both maximum displacement and maximum throw. Narrow confidence intervals and relatively high coefficients of determination suggest coherent deformation and limited structural interference. These features imply that faults in these domains developed under relatively stable stress conditions and possess similar mechanical properties. This behaviour aligns with fault populations that have advanced beyond early segmentation into more laterally continuous structures (Fossen & Rørnes, 1996; Marrett & Allmendinger, 1991, 1992).

In contrast, fault domains FM-3 and FM-4 exhibit weaker correlations and greater scatter in their scaling relationships. Broad confidence intervals and diminished regression fits suggest heterogeneous deformation and uneven strain distribution. These domains feature faults influenced by segmentation, linkage, and potential reactivation, where local geological factors override idealised scaling behaviour. Similar deviations from classical scaling relationships have been observed in structurally complex basins and in systems affected by mechanical layering or ductile detachments (Baudon & Cartwright, 2008; Gudmundsson et al., 2013).

Displacement–distance profiles offer additional insights into fault evolution. Mature faults usually display smooth, bell-shaped displacement profiles, with peak displacement near the fault centre and decreasing towards the tips, indicating efficient strain distribution (Dawers et al., 1993; Marrett & Allmendinger, 1992). In contrast, segmented or immature faults show irregular profiles with multiple displacement peaks, suggesting incomplete linkage and ongoing fault growth. The profiles observed cover this entire spectrum, supporting the idea of progressive fault development across the mapped regions.

Displacement gradient analysis offers an additional quantitative measure of fault maturity and strain localisation. By normalising the maximum displacement relative to half the fault length, displacement gradients reveal the rate at which displacement decreases along the fault plane. Higher gradients suggest strongly localised slip and immature or reactivated faults, while lower gradients indicate more mature faults with distributed deformation (Fossen & Rørnes, 1996; Gillespie et al., 1993; Marrett & Allmendinger, 1992). The broad range of displacement gradients observed, especially in FM-4, emphasises the impact of structural complexity and multiphase deformation on fault behaviour.

Notably, the weak correlation between displacement gradient and fault length observed highlights the limitations of relying on a single quantitative parameter. Fault growth does not follow a single evolutionary pathway; instead, it reflects the combined effects of stress history, fault interaction, mechanical stratigraphy, and boundary conditions. Quantitative metrics, therefore, have their most significant value when interpreted alongside structural style, fault geometry, and stratigraphic setting.

11.7 Comparative Distribution of the Levantine Margin and Other Regions of the World

A comparative analysis of the Levantine Margin, the offshore western Niger Delta, Orphan Basin–Flemish Pass, and the North Chukchi Basin–Hanna Trough reveals that rift architecture is mainly governed by mechanical stratigraphy and detachment efficiency, rather than by the magnitude of extension alone. Throughout all margins, extensional strain is absorbed by normal faulting, although fault geometry, density, polarity, and interaction style differ depending on the depth, rheology, and continuity of mechanically weak layers.

Margins characterised by shallow, weak detachments, such as Messinian evaporites on the Levantine Margin or mobile shales in the Niger Delta, develop dense, segmented fault networks with a mix of synthetic and antithetic polarities. These systems show complex fault interactions, including relay ramps, splays, and conjugate geometries, which reflect significant stratigraphic decoupling and local stress reorganisation (Brun & Choukroune, 1983; Buck, 1991; Morley & Guerin, 1996). In contrast, margins without effective shallow detachments, such as the North Chukchi Basin–Hanna Trough, localise deformation onto fewer, steep, basement-involved faults with stable polarity and vertically coherent displacement profiles (Kumar et al., 2011; MMS, 2006).

Fault polarity is a sensitive indicator of strain partitioning and mechanical coupling. Detachment-controlled systems show rapidly alternating polarities and conjugate fault patterns, whereas more strongly coupled systems maintain consistent polarity across extensive structural domains. These polarity patterns do not represent tectonic inversion but instead record displacement compatibility above weak detachments and evolving stress fields during fault growth (Morley and Guerin; Cawood et al., 2021; Chima et al., 2022).

Kinematic analysis further shows that fault interaction styles vary with detachment efficiency. Distributed systems favour relay-rich, adaptive fault growth, while detachment-dominated or basement-involved systems encourage early linkage and efficient strain transfer along fewer structures. Acute bisector relationships consistently indicate extension-dominated deformation, confirming that intersection geometry is a strong proxy for regional strain direction, even in structurally complex fault networks.

Together, these observations outline a spectrum of extensional behaviour, from distributed, detachment-controlled margins to localised, basement-involved rifting. Notably, margins can shift along this spectrum over time as sedimentation rate, detachment strength, or extension magnitude change. This framework offers a process-based approach for interpreting fault networks, predicting subsurface connectivity, and differentiating gravity-driven deformation from crustal stretching on rifted continental margins.

11.8 Scaling Fault System Evolution from Analogue Models to Seismic Data

Analogue sandbox experiments and 3D seismic interpretation together offer a complementary, multi-scale framework for understanding the development of extensional fault systems. This synthesis demonstrates that analogue models effectively replicate the essential mechanical processes of fault nucleation, growth, interaction, and linkage. Meanwhile, seismic data record the time-integrated structural complexity inherent in natural rift systems.

Across scales, both approaches consistently generate planar and listric normal faults, rotated fault blocks, graben–horst architectures, and relay structures, confirming that extensional deformation adheres to robust mechanical rules regardless of scale (McClay, 1990; McClay & Ellis, 1987; McClay et al., 2002). Analogue models typically depict early-stage deformation using uniform materials and controlled boundary conditions, resulting in regular fault spacing, symmetrical grabens, and rapid development of kinematically efficient fault systems. The Levantine Margin shows similar first-order geometries but with greater variability in fault spacing, asymmetry, and curvature due to stratigraphic heterogeneity, evaporitic detachments, and inherited structures.

Fault polarity provides a key link between experimental and natural systems. Sandbox experiments show, extension-parallel polarity patterns dominated by synthetic fault arrays (McClay & Ellis, 1987; McClay et al., 2002). By contrast, seismic data preserve mixed-polarity configurations, including synthetic–synthetic, synthetic–antithetic, and antithetic–antithetic relationships. This diversity reflects mechanical decoupling within layered stratigraphy and progressive stress reorganisation during fault growth. Significantly, polarity complexity decreases as faults mature, mirroring the evolutionary simplification observed in analogue end-member systems.

Kinematic interactions further differentiate scale effects. Analogue models emphasise efficient relay development and early fault linkage once critical overlap thresholds are exceeded (McClay et al., 2002; Wood, 1988). Seismic-scale fault networks retain a wider range of interaction styles, including non-breached relays, transient linkage geometries, splays, and cross-cutting relationships, reflecting prolonged deformation histories and evolving mechanical conditions. Nevertheless, acute bisector relationships in both systems consistently indicate extension-dominated strain, confirming that intersection geometry provides a reliable proxy for deformation direction across scales.

This comparison demonstrates that analogue models function best as mechanical templates rather than exact geometric replicas. They represent the fundamental physics and end-members of fault evolution, whereas seismic data record the cumulative effects of time, stratigraphic layering, and structural inheritance. When combined, analogue insights assist seismic interpretation by clarifying expected fault behaviour, while seismic observations ground analogue predictions in geological reality.

11.9 Implications

Three-dimensional (3D) seismic interpretation of fault interactions and linkages has transformed the understanding, modelling, and prediction of geological and tectonic processes, structural deformation, and subsurface architecture. This advanced method has important implications across disciplines such as geology, geophysics, engineering, and resource exploration. Here are some key implications of 3D seismic interpretation of fault interactions and linkages:

1. Accurate Structural Understanding: Traditional two-dimensional seismic interpretations offer limited insight into fault systems, often oversimplifying them. 3D seismic data provide a comprehensive view of fault networks, enabling geoscientists to accurately map fault geometries, orientations, and interactions in the subsurface. This interpretation improves structural understanding, which is essential for constructing precise geological models.
2. Fault Propagation and Evolution: 3D seismic data allow researchers to analyse how faults propagate and develop. By visualising fault interactions in three dimensions, geoscientists can reconstruct the kinematic history of fault systems, shedding light on fault initiation, growth, and the migration of deformation.
3. Strain Accommodation: Detailed 3D seismic interpretations offer a clearer understanding of how faults interact to accommodate strain in the Earth's crust. This knowledge is crucial for determining stress distributions within fault networks and for evaluating seismic hazard and risk.
4. Reservoir Compartmentalisation: It is a crucial aspect of subsurface exploration and production, especially in hydrocarbon reservoirs. Fault barriers can limit fluid communication between fault blocks, thereby impacting sweep efficiency during production. Conversely, fault-related compartmentalisation can create

isolated hydrocarbon pockets, leading to missed production opportunities if not properly identified and managed.

5. Reservoir Connectivity: Understanding the connection between fault systems and reservoirs is essential for efficient hydrocarbon recovery. 3D seismic interpretations allow geoscientists to identify fault corridors that support fluid flow, thereby improving our understanding of subsurface fluid dynamics and reservoir performance.
6. Fracture Networks in Geothermal Systems: Geothermal energy production depends on permeable fracture networks. 3D seismic data help locate and characterise these fractures, offering insights into the pathways through which fluids circulate, and heat is extracted from the Earth's crust.
7. Enhanced Engineering Design: The understanding of fault interactions obtained via 3D seismic interpretation is crucial for engineering design. This knowledge is especially pertinent for tunnels and underground storage facilities.
8. Infrastructure development projects: Engineers can consider potential ground instability caused by fault movements and design structures to withstand these challenges.
9. Earthquake Hazard Assessment: 3D seismic interpretation improves earthquake hazard assessments by uncovering complex fault systems. Understanding fault interactions and linkages helps geoscientists identify potential earthquake sources, assess rupture scenarios, and estimate ground-shaking intensities across various fault configurations.
10. Induced Seismicity Studies: Hydraulic fracturing and reservoir stimulation can cause seismic events. 3D seismic interpretations assist researchers in identifying existing faults and their potential interactions with induced fractures, thereby supporting the assessment and mitigation of induced seismicity risks.
11. Environmental Impact Assessments: The environmental impact of activities such as mining, construction, and energy production can be evaluated using 3D seismic interpretations. Understanding the subsurface fault network aids in assessing risks like surface subsidence, groundwater contamination, and induced seismicity.
12. Climate and Landscape Evolution: Fault systems influence landscape development and sedimentary deposit distribution. 3D seismic interpretations reveal how fault interactions affect sediment transport, basin formation, and the shaping of Earth's surface over geological timescales.
13. Educational and Outreach Opportunities: The visual aspect of 3D seismic interpretations enhances their value for education and public engagement. Interactive displays and simulations based on these interpretations can assist in explaining complex geological concepts to students, professionals, and the public.
14. Carbon Capture Utilisation and Storage (CCUS): Conducting a 3D seismic interpretation of fault interactions and linkages in CCUS has multiple implications, including improved reservoir characterisation, risk assessment, containment assurance, site selection, injection strategy, monitoring design, optimised well placement, regulatory compliance, and long-term monitoring. Collectively,

these aspects support the safe and effective realisation of CCUS projects to reduce CO2 emissions and combat climate change.

11.10 Future Guidance and Methodological Advances

Seismic interpretation continues to advance as improvements in data quality, computational power, and analytical methods transform how geoscientists characterise the subsurface. It emphasises several methodological directions that can further enhance the accuracy, transparency, and usefulness of seismic interpretation, especially in structurally complex environments.

One key future direction is integrating probabilistic, uncertainty-aware workflows. Traditional deterministic interpretations produce a single "best-fit" structural model, but seismic data often support multiple plausible geological scenarios. Recent studies promote probabilistic depth conversion, stochastic fault placement, and ensemble-based interpretation to explicitly represent uncertainty ranges rather than single solutions. Embedding these approaches within structural–stratigraphic frameworks would allow interpreters to quantify confidence in fault geometry, displacement, and connectivity.

Another significant development is the increasing role of machine learning and artificial intelligence in seismic interpretation. Automated fault detection, horizon tracking, and attribute extraction algorithms can greatly reduce interpretation time and enhance consistency across large datasets. However, this approach emphasises that such tools must function within geologically constrained frameworks. Without a stratigraphic and structural framework, automated methods risk amplifying noise, artefacts, or interpreter bias present in the training data. Hybrid workflows that combine machine learning with rule-based geological validation present the most promising way forward.

Improvements in seismic imaging and resolution will continue to influence interpretation strategies. Advances in broadband acquisition, depth imaging, and velocity model building enhance the detection of subtle structures and small-offset faults. However, modelling studies demonstrate that resolution limits will always impose fundamental constraints on what seismic data can resolve. Future interpretation workflows must therefore continue to emphasise resolution-aware fault detection criteria and the transparent communication of detection thresholds.

The integration of multidisciplinary data represents another vital advancement. Where accessible, well logs, rock physics models, and geomechanical data offer crucial constraints on velocity, lithology, and fault behaviour. Combining seismic interpretation with geomechanical modelling can enhance predictions of fault reactivation, seal integrity, and the potential for induced seismicity, especially for CCUS and geothermal applications.

Equally important is the increasing focus on reproducibility and interpretive transparency. As subsurface models more frequently influence regulatory decisions and public-facing energy projects, interpreters must clearly document assumptions, workflows, and sources of uncertainty. Reproducible interpretation practices, such as version-controlled models, interpretable decision criteria, and shared uncertainty metrics, are becoming vital to best practice in subsurface geoscience.

Finally, future research should continue to bridge the gap between conceptual geological understanding and quantitative analysis. Fault scaling relationships, displacement gradients, and statistical metrics are powerful tools, yet their interpretation remains setting-dependent. Progress will come not from replacing geological reasoning with automation, but from integrating quantitative diagnostics into robust geological frameworks that recognise complexity and uncertainty.

References

Alcalde, J., & Bond, C. E. (2022). Subjective uncertainty and biases: The impact on seismic data interpretation. In *Interpreting subsurface seismic data*. Elsevier, pp. 103–123.

Andrews, B. J., Mildon, Z. K., Jackson, C.A.-L., & Bond, C. E. (2024). Quantifying fault interpretation uncertainties and their impact on fault seal and seismic hazard analysis. *Journal of Structural Geology, 184*, Article 105158.

Bacon, M., Simm, R., & Redshaw, T. (2003). *3D seismic interpretation*. Cambridge University Press.

Baudon, C., & Cartwright, J. (2008). Early-stage evolution of growth faults: 3D seismic insights from the Levant Basin. *Journal of Structural Geology, 30*, 888–898.

Bond, C. E. (2015). Uncertainty in structural interpretation: Lessons to be learnt. *Journal of Structural Geology, 74*, 185–200.

Bond, C. E., Gibbs, A. D., Shipton, Z. K., & Jones, S. (2007). Conceptual uncertainty in geoscience interpretation. *GSA Today, 17*, 4–10.

Brown, A. R. (2012). Interpretation of three-dimensional seismic data, 6th edn. AAPG/SEG.

Brun, J.-P., & Choukroune, P. (1983). Normal faulting, block tilting and décollement in a stretched crust. *Tectonics, 2*, 345–356.

Buck, W. R. (1991). Modes of continental lithospheric extension. *Journal of Geophysical Research, 96*, 20161–20178.

Cartwright, J. A., Mansfield, C. S., & Trudgill, B. (1996). The growth of normal faults by segment linkage. *Geological Society, London, Special Publication, 99*, 163–177.

Carvajal, F. R., Butler, R. W. H., & Bond, C. E. (2023). Mapping faults in 3D seismic data—why the method matters. *Journal of Structural Geology, 177*, Article 104976.

Cawood, A. J., Ferrill, D. A., Morris, A. P., Norris, D., McCallum, D., Gillis, E., & Smart, K. J. (2021). Tectonostratigraphic evolution of the Orphan Basin and Flemish Pass region. *Marine Petroleum and Geology, 128*, 105042; *133*, 105219.

Chima, K. I., Granjeon, D., Couto, D. D., Leroux, E., Gorini, C., Rabineau, M., & Letouzey, J. (2022). Tectono-stratigraphic evolution of the offshore western Niger Delta. *Basin Research, 34*, 25–49.

Dawers, N. H., Anders, M. H., & Scholz, C. H. (1993). Growth of normal faults: Displacement–length scaling. *Geology, 21*, 1107–1110.

Dimmen, V., Rotevatn, A., & Lecomte, I. (2023). Imaging of small-scale faults: Insights from seismic modelling. *Marine and Petroleum Geology, 147*, Article 105980.

Faleide, T. S., Braathen, A., Lecomte, I., Mulrooney, M., Midtkandal, I., Bugge, A. J., & Planke, S. (2021). Impacts of seismic resolution on fault interpretation: Insights from seismic modelling. *Tectonophysics, 816*, Article 229008.

Fossen, H., & Rørnes, A. (1996). Properties of fault populations in the Gullfaks Field, northern North Sea. *Journal of Structural Geology, 18*, 179–190.

Gibbs, A. D. (1984). Structural evolution of extensional basin margins. *Journal of the Geological Society of London, 141*, 609–620.
Gillespie, P. A., Howard, C. B., Walsh, J. J., & Watterson, J. (1993). Measurement and scaling of spatial distributions of fractures. *Tectonophysics, 226*, 113–141.
Gudmundsson, A., De Guidi, G., & Scudero, S. (2013). Length–displacement scaling and fault growth. *Tectonophysics, 608*, 1089–1100.
Kumar, N., Granath, J. W., Emmet, P. A., Helwig, J. A., & Dinkelman, M. G. (2011). Stratigraphic and tectonic framework of the U.S. Chukchi Shelf. *Geological Society, London, Memoirs, 35*, 501–508.
Lowell, J. D. (1985). Structural styles in petroleum geology. OGCI, Tulsa, OK
Marrett, R., & Allmendinger, R. W. (1991). Estimates of strain due to brittle faulting. *Journal of Structural Geology, 13*, 735–738.
Marrett, R., & Allmendinger, R. W. (1992). Amount of extension on "small" faults: An example from the Viking Graben. *Geology, 20*, 47–50.
McClay, K. R. (1990). Extensional fault systems in sedimentary basins: A review of analogue model studies. *Marine and Petroleum Geology, 7*, 206–233.
McClay, K. R., & Ellis, P. G. (1987). Geometries of extensional fault systems developed in model experiments. *Geology, 15*, 341–344.
McClay, K. R., Dooley, T., Whitehouse, P., & Mills, M. (2002). 4-D evolution of rift systems: Insights from analogue modelling. *AAPG Bulletin, 86*, 935–959.
Michie, E. A. H., Mulrooney, M. J., & Braathen, A. (2021). Fault interpretation uncertainties using seismic data and the effects on fault seal analysis. *Solid Earth, 12*, 1259–1286.
Minerals Management Service. (MMS). (2006). Chukchi sea planning area—province summary. MMS, Anchorage, AK, USA
Morley, C. K., & Guerin, G. (1996). Comparison of gravity-driven deformation styles associated with mobile shales and salt. *Tectonics, 15*, 1154–1170.
Needham, T., Yielding, G., & Fox, R. (1996). Fault population description and prediction. *Journal of Structural Geology, 18*, 155–167.
Onajite, E. J. (2014). *Seismic data analysis techniques in hydrocarbon exploration*. Elsevier.
Pérez-Díaz, L., Alcalde, J., & Bond, C. E. (2020). Handling uncertainty in the geosciences: Identification, mitigation, and communication. *Solid Earth, 11*, 889–897.
Shelton, J. W. (1984). Listric normal faults: An illustrated summary. *AAPG Bulletin, 68*, 801–815.
Steventon, M. J., Jackson, C.A.-L., Hall, M., Ireland, M., Munafo, M., & Roberts, K. (2022). Reproducibility in subsurface geoscience. *Earth Science, Systems and Society, 2*, 10051.
Stewart, S. A. (2011). Vertical exaggeration in seismic interpretation. *Marine and Petroleum Geology, 28*, 959–965.
Stewart, S. A. (2012). Interpretation validation on vertically exaggerated reflection seismic sections. *Journal of Structural Geology, 41*, 38–46.
Stone, D. S. (1991). Analysis of scale exaggeration on seismic profiles. *AAPG Bulletin, 75*, 1161–1177.
Tannert, C., Elvers, H.-D., & Jandrig, B. (2007). The ethics of uncertainty. *EMBO Reports, 8*, 885–889.
Widess, M. B. (1973). How thin is a thin bed? *Geophysics, 38*, 1176–1180.
Wood, R. (1988). Relays, bends, and accommodation zones in standard fault systems. *Geological Society of London Special Publication, 28*, 311–325.

MIX
Papier aus verantwortungsvollen Quellen
Paper from responsible sources
FSC® C105338

If you have any concerns about our products,
you can contact us on
ProductSafety@springernature.com

In case Publisher is established outside the EU,
the EU authorized representative is:
Springer Nature Customer Service Center GmbH
Europaplatz 3, 69115 Heidelberg, Germany

Printed by Libri Plureos GmbH
in Hamburg, Germany